The Search for Human Chromosomes

Wilson John Wall

The Search for Human Chromosomes

A History of Discovery

 Springer

Wilson John Wall
Bewdley
United Kingdom

ISBN 978-3-319-26334-2   ISBN 978-3-319-26336-6   (eBook)
DOI 10.1007/978-3-319-26336-6

Library of Congress Control Number: 2015958021

Springer Cham Heidelberg New York Dordrecht London
© Springer International Publishing Switzerland 2016

Springer International Publishing AG Switzerland is part of Springer Science+Business Media
(www.springer.com)

Introduction

Why do we need to know how many chromosomes there are in a cell. What are chromosomes made of and how can their behaviour have far-reaching effects on inheritance and health? Simply put—why are they so important and why should we all know about them?

There comes a point in the curiosity of man when simple observation and description is not enough. This was the point at which genetics was born from observations of plant and animal breeding. The problem of most breeding is that to a casual observer it seems to produce random results, which can then be refined by inbreeding. More than that, some species have a distinct propensity to remain unchained by breeding, whereas species such as dogs can be changed out of all proportion by breeding. This variation in levels of plasticity of the genome in some species gives rise to far greater variation between breeds than in species that do not seem to change to such a large degree. Some species are reluctant producers of varieties for different reasons, for example, large and long-lived trees are not good experimental organisms because they will most likely outlive their researcher before any useful changes can be observed. It may also be that there may not be any reason for a breeder to cast an eye on an already useful and naturally regenerating species.

We all know that the coming of the human genome mapping project was seminal in the development of our perception of ourselves, but what most don't realize is that a far more technical and intellectually demanding process takes place in genetic laboratories around the world every day. This is the counting and quantifying of chromosomes. You see, we know that normal human body cells, nonreproductive cells other than red blood cells, all have a complement of 46 chromosomes, but within that is a world of control and expression which depends not just on that number but also on how the material is distributed amongst the chromosomes. Some chromosomes can swap material between them without problem, and some can stick together and cause no trouble for the individual, but may cause untold damage to future generations. Much of this we can now detect and advise on, some can only be detected when the damage has started, as in some cancers as not all tumour-causing damage is due solely to single gene mutations. Chromosomes are complicated. We are looking at the similarity between having a dictionary and saying all the words are there—it must be the complete works of Shakespeare. The words have to be in the right order, sentences have to be started and stopped at the right points, and the players must know their marks. Knowing the human genome is as nothing to knowing how chromosomes structure and control the genome, allowing levels of expression as appropriate in different tissues.

All the knowledge which we have about chromosomes and their importance started with the understanding of three things: the first is that they are constant in number, the second is that they are complex structures, and the third is that these are the carriers of the genes. It is surprising that the second and third of these were quickly understood, while the first 'that they are constant' was realized, but the number was for a long time tantalizingly out of reach. In fact for many years, the human chromosome number was erroneously thought to be 48, rather than the true 46. Actually, we glibly say it is 46, but this is really the modal number as cells in culture regularly loose chromosomes. This is not so strange as a cell in culture does not need anywhere the full complement of genes that a tissue or complete organism does.

We know that a chromosome carries all the genes, but it is more than that; there is a way in which the chromosomes carry genes, but only in the same way that sleepers carry railway track. They are an integral part of the structure. Without the sleepers the track is just steel; without the genes the chromosome is just a mixture of protein and associated chemicals. At the same time, from a philosophical point of view, the genes are the content, the nebulous part of the chromosome which has no solid existence until they are transcribed into their functional products. The genes on a chromosome are no more than the water molecules in a wave—until the wave hits the shore and throws flotsam onto dry land.

Bewdley   Wilson John Wall

Contents

Background to the Hunt for the Human Chromosome Number

1

When we think of biology in the twenty-first century, it is hard to understand all we know of genetics in any sound form we have gained in less than two centuries—before that it was nearly all speculation. It is little more than a century since the simple rules of inheritance were discovered. This is the generation in which the structure of DNA was conclusively demonstrated, and although it may seem odd, that discovery predates the discovery of the human chromosome number by a few years. This is probably because although the crystallographic data of DNA took a lot of interpretation, finding the number of human chromosomes was a far more technically complicated problem. Even so, the structure of DNA and chromosome number were all worked out within a single lifetime.

A question which is often asked by biochemists is why we need to know anything at all about chromosomes. The argument runs that if we know the sequence of a gene, we know everything that we need to know about that gene. This is a question never asked by biologists and geneticists who understand the complexity of biology and the genetic legacy we all carry. Chromosomes are part of this complexity. They are complicated structures carrying the genes and creating complexity in the way they allow and control gene expression.

Look at it another way. If you have a diagnostic procedure that involves cytogenetics, the study of chromosomes, you are on the receiving end of the most complicated and technically demanding medical test that is currently available. This is why we need to know as much about human chromosomes as we can, the number structure and function. Hidden from view for a long time, they continually reveal their influence on our genetic makeup. It is worth remembering that a chromosome, any chromosome, either plant or animal, is made up of a single molecule of DNA mixed up with proteins of various sorts. In this way chromosomes can be seen as huge molecules, too big to handle chemically, too small to be seen with the naked eye. This is a recipe for problems, which would take a lot of sorting out.

It is perhaps because of this rather unusual position that chromosomes hold in the story of inheritance that it was after much was known about the chemistry of inheritance that the human chromosome number was discovered. What is even

© Springer International Publishing Switzerland 2016
W.J. Wall, *The Search for Human Chromosomes*,
DOI 10.1007/978-3-319-26336-6_1

more remarkable is that it is not what you would imagine as a logical order of discovery that took place. We would expect to move from the large, chromosomes, to the small, the molecule, but it transpired to be the other way round. So first it was DNA as the molecule of inheritance, next the structure of DNA and only finally the human chromosome number. There is a simple explanation for this; partly it is that chromosomes, as we shall see, can be extremely difficult to handle. But also the other two aspects of DNA are questions of chemistry. The answers do not depend upon the quirks of biology—the study is simple in comparison because it yields easily to a reductionist scientific method.

There is an idea that a reductionist view of the world both helped with the progress of biological thought and hindered it. It helped by providing simple answers to simple questions, but it hindered when faced with questions that had no simple answer—like why we have five fingers.

We inherited this reductionist view of the natural world from previous centuries. It works well in physics and chemistry, but does not always yield answers in biology. The notion that we could reduce everything to a single answer came down to us from thinkers based around the Mediterranean. That means the Greeks and Romans mainly. They were, for the most part, working on thought experiments. As they had little or no equipment to play with, the way it worked was to pose a question which was then answered with the simplest explanation they could muster. If this fitted the observations, then it was regarded as true, it being unnecessary to make the explanation more complicated. As a method of thinking, this works very well, but only if the observations are complete, and in the case of biology, this was not often the case.

What these thinkers did 2500 years ago was to reduce the supernatural component as far as possible while still trying to explain what they observed. Indeed, Hippocrates stood firmly by this method when describing disease conditions. As far as he was concerned, disease was entirely organic. This is well demonstrated by his straightforward assertion that there was no supernatural component in epilepsy. At the time of his writing about the condition, epilepsy was commonly called the sacred disease because it was said that the sufferer was possessed by a god while having an attack. Note that it was possession by a god, rather than God, as they had a polytheistic religious system. He went further saying that there was a natural cause of epilepsy, and with some insight that even if we did not know what it was, to suggest a godly content of the disease was invoking superstition to cover ignorance. The biological speculations of people such as Hippocrates were inevitably based around medical matters. Then, as now, medicine was regarded as the most important application of biology because sickness has stalked humanity throughout the ages, while the more interesting motivation of curiosity requires a different economy and social structure (Fig. 1.1).

Of all these ancient philosophers, for many people the one who is best known is Aristotle. Mainly regarded as a philosopher and writer, he has another claim as being the father of the life sciences. The difference between the philosopher physicians that both predated and came after Aristotle was that they were interested in biology more as an abstract concept than a material reality. Aristotle, on the other hand, was a meticulous observer of the natural world in all its forms, trying to explain what he saw as clearly as possible (Fig. 1.2).

Fig. 1.1 Hippocrates: this statue is from Rangaraya Medical College, India, and reflects the international legacy of Hippocrates

Fig. 1.2 Bust of Aristotle, a Roman copy in marble of a Greek bronze by Lysippus. 330 BC

Asking difficult questions, which for the most part they had no chance of answering, they inevitably simplified both the question and the answer. This was the formation of reductionist thinking which dominated following centuries. One of the questions that they asked was on the origins of life, a question where there is still no consensus. To philosophers of the age, lack of any significant data would not hinder speculation; the mind should be able to answer such questions starting from the most basic of principles. One answer to this conundrum was put forward by Anaximander in about 520 BC in a suggestion that covered both cosmology and zoology. It is often said that his was an idea which predated evolution, but it was both simpler and more pragmatic than that. He suggested that the first living things were generated out of primeval slime by the heat of the sun, and later species emerged out of prickly husks onto the dry land. This may be taken as evolving, but only in the broadest sense. His was much more a description of progressive

Fig. 1.3 Anaximander.
Probably a Roman copy of a
Greek original relief (Museo
Nazionale Romano)

Fig. 1.4 Hendrik ter
Brugghen painting of
Democritus (Rijksmuseum
Amsterdam)

creation, culminating in man. As an idea it certainly gave no hint of any mechanism
of inheritance; in fact it did away with consistent species altogether (Fig. 1.3).

Less than a hundred years after Anaximander, a group of philosophers led by
Leucippus and Democritus came up with another often-quoted hypothesis, atom-
ism. Sometimes considered to be the origins of the atomic theory of matter, in
reality it has a much broader brush stroke than that. The theory started from the
premise that nothing comes from, or is reduced to, nothing. The void is infinite and
so is the number of atoms, which come in all different shapes and sizes, although
none of them can be seen as they are too small. By this philosophy, there is nothing
else, so everything has to be explained in terms of their atoms. So a thought
experiment explained everything about the world. But it heralded something
more, the far more important idea that some things could not be seen but were
not supernatural (Fig. 1.4).

Now, we have said that Aristotle was the founder, the keystone, of biology. He
also denied the ideas of the atomists. This may seem strange to us, but while

Aristotle was a brilliant observer of the plants and animals around him, the atomists had no measure or experiment to back up their ideas, only logical inference. From this it is reasonable that Aristotle should say that what he saw he knew and understood; there was no need for the atomist ideas of progressively smaller and smaller, invisible particles. What Aristotle brought to biology was at the time unique. He had an unprecedented ability to observe and at the same time interpret what he saw. We shall see later that throughout the search for the human chromosome number and throughout biology as a whole, observation is easy but interpretation is not. It is this fundamental method of reasoning which made his concepts of taxonomy and morphology, especially in embryology, which mark him out so clearly.

As biologists there is much for us to admire in his work; he described 500 animals and put them into eight different classes, clearly demonstrating that he understood that there were relationships between the various animals. With his attention to ecological problems, Aristotle also seemed to have had a clear understanding of the overall importance of interactions between species. For Aristotle every species was complete; there was no concept of evolution and yet species could be arranged in order of seniority from the lowest and simplest to the highest and most complex. While his ideas on biological form and function contain rather a large chunk of teleology for modern tastes, there is no gainsaying his incredible ability to observe and put his observations into context. While mentioning the teleological aspects of his ideas, it is worth remembering that it still crops up in modern conversation with statements such as 'it will adapt to a new environment' as though something is guiding the process. We may not fully understand the implications of what we say, but embedded in that simple statement is the assumption that the organism has some control over something about which it is completely unaware. This is a teleological fallacy.

It is the legacy of observing and explaining which Aristotle left behind that has endured. But it was his work on taxonomy which fuelled the earliest ideas about genetics and inheritance. Some notions were slightly off beam, or in some cases very off beam, but they all started with one broad question. Why do species breed true? Why did horses give rise to horses and chickens give rise to chickens? Basically, why do species only regenerate themselves? This basic question was going to echo down the centuries. It was not going to be provided with an answer until well into the nineteenth century by which time the basics of biology would be understood. It would actually be even longer before chromosomes would be involved in answering such questions. Inheritance was going to furnish an excellent example of a theoretical explanation being quite separate from a tangible one. Before anything else regarding inheritance, we would see a mathematical demonstration of the mechanism of inheritance without any explanation of how it was done.

The reasons for this strange about face set of discoveries, a theoretical explanation before a physical one, were manifold, but do stem from the reductionist view of the world generated by physics and chemistry. These two subjects tend towards a linear explanation of the universe, striving for easily encapsulated simple rules and formulae. As a consequence, biology in general and genetics in particular were also seen as needing simple rules to govern it. With hindsight, if we took this as true, it would rule out the chromosomes as being of any consequence other than as physical vehicles for genes. Genes, in the form of a DNA sequence, would be the lynch pin, and no more information would be required to define an organism and all that it does. This, of course, is not the case. Not only are chromosomes the carriers of the information, they have a complicated structure which helps and controls gene expression. Besides this, we now know the number of chromosomes is crucial to a species. Also the position of genes on chromosomes is important. Before any of this could be realised, basic cellular structure and function had to be worked out, which was going to be a massive undertaking in itself.

It was long after Aristotle had laid down ideas about empirical data that the first clear visualisation of cells was made. This was hardly surprising as observation with the naked eye would reveal nothing—equipment had to be invented. But since people struggling with these problems did not know what questions to ask, it would be the equipment often designed for different purposes that would be brought to bear on the task. Nowadays, we think of cells as being the fundamental unit of life, but until they were seen such things were not even thought to exist. It was Robert Hooke, using one of his own microscopes, who is considered to be the original observer of cells, and it is to him that we also owe the word cell as designating, well, cells. This is not just significant for biology, but it was also closing the gap between what was imagined and what could be seen in the physical world about us.

Hooke had coined the word cell for use in biology in 1665 (Hooke 1665), although dictionaries quote the date as 1672. It was from this point in the seventeenth century that it was known that cells exist and that in some magical way, perhaps even a divine way, a species gave rise to copies of itself. For this uncritical attitude to reproduction of living things, we have the church to thank. The immutability of species was all to do with the hubris of complacent religions which claimed to know the answers, infallibly. This was so even when their versions of truth were no better than reading a chicken's entrails as a way to determine tomorrow's weather. To make real headway and crack the inheritance question, not to answer it, but simply ask it in a coherent manner that could eventually be answered, a clear idea of cells and organisms was required. In the early nineteenth century, two remarkable figures entered this arena.

It was about 1830 and two friends, working independently and on different material, constructed what today we call the cell theory. This is simply that all tissues and organisms are made up of cells. You see, until they formulated the idea, it was assumed that organisms were in some way a whole and each organ was a complete unit, no cells involved. This was quite understandable since tissues and organs look different from each other and yet have a completeness unto themselves.

The two scientists who came up with the idea were Theodor Schwann and Matthias Jakob Schleiden. Schleiden was a botanist with extensive studies of plant material under his belt which led him to make accurate deductions that cells were both universal and of primary importance. Schwann, on the other hand, was a biologist with a specific interest in animal rather than plant material. Having trained in physiology, he came up with a description of myelinated nerve cells that eventually became known as the Schwann cell. Besides this he described the single-celled nature of the egg and managed to demonstrate that spontaneous generation of life was impossible. This latter feat was probably his greatest demonstration but was not universally accepted until Pasteur repeated his findings much later in the early 1860s. In his later years he moved from Germany to Belgium where he more or less gave up science. Schwann clearly defined their cell theory as they understood it in his publication of 1839 (Schwann 1839).

There was one flaw in the Schleiden and Schwann cell theory in that they thought that new cells arose from a budding process and not by cells dividing into equal halves. This is a minor point, really, because they had no idea how inheritance worked, what makes that particular type of species retain its identity. They would have guessed it had to be passed on in some physical form, but did not even speculate how this could be done. There was still a retained idea that the cell was as it was, and therefore all the daughter cells would follow suit; philosophically they could not change. This was in the same way that a piece of string when cut in half is now two pieces of string; no information has to be passed on for the string to retain its identity.

The second half of the nineteenth century saw considerable leaps forward in biology, specifically genetics and inheritance. This was because like all good science, questions were being asked that had to be answered, without recourse to vague notions of untestable determining forces. Increases in understanding of what was and still is very small components of the cell took the form both of a theoretical framework and a practical, experimental framework. Pivotal to the changes in attitude to inheritance was the publication in 1858 of papers by Charles Darwin and Alfred Russel Wallace (Darwin and Wallace 1858). These were developed and encapsulated a year later in *The Origin of Species by Means of Natural Selection* (Darwin 1859).

Although evolution is all to do with inheritance and ultimately genetics, it stands to genetics in the same way as epidemiology does to human disease. It deals with the overall picture, not with the minutiae of how and why it takes pace and what it means to the individual. The major point of criticism of Darwin's ideas at the time was that there was no understanding of the mechanism, whereby variation and heredity could take place and be inherited. This was a problem because explicit within the work of Darwin was the requirement for adaptations to be inheritable. Without the inheritability of adaptive characters, there would be no evolution. What was happening during this period of theoretical development in the nineteenth century was the restatement of observed events as if they were self-explanatory. There were no experimental results which followed any sort of comprehensible pattern. All that could be said was that there must be a mechanism of inheritance because inheritance requires there to be so. This process of restating the observed phenomena in a different way as though it was an explanation is a common cover for lack of knowledge.

It was only a very few years after Darwin had published his groundbreaking work that results were published which filled the gap between vague notions of heredity and specific ideas of inheritance. It was a more specific piece of genetics than had ever been published before as it had data to back up its claims. In 1865 Gregor Mendel, an Augustinian monk, delivered a series of papers to the Brunn Natural History Society. These were collected and published by the society in 1866 (Mendel 1866). This is when genetics as we know it started. The implications were profound and so were the lines of reasoning which culminated in the experiments that changed our view of inheritance.

Gregor Mendel had an unlikely start both as a monk and a scientist. He joined the monastery at the age of 21 in 1848. He also tried twice to pass the teaching exams of the order but failed both times. His plant breeding, which he carried out in the monastery garden, was more or less abandoned after he was elected abbot in 1868. It was many years after his death in 1884 that the precise implications and importance of his work was appreciated. So important is the work of Mendel that is dealt with in far more detail in Chap. 3. It is here that the basic results and the way they are interpreted are explained.

Part of the complexity of demonstrating discrete inheritance was that it had always been assumed that characters were 'blended' between generations. The idea that inheritance might be guided by discrete factors seemed both too complicated to understand and unnecessarily complicated if it was to explain certain observations. Here again the old reductionist view of blending characters in some magical way was seen as the simplest and therefore most likely explanation. For example, a child might look like the parents, but not exactly; the child to all intents and purposes is a blend of the parents. Also some conditions were known to skip generations, which can be difficult to explain unless some idea of dominant and recessive traits is used, thereby creating unnecessary complexity. There is a more subtle contradiction going on here, though. If characters are blended between generations, then how could a small and subtle advantage be passed on to future generations without being lost? This process is essential for evolution to work.

At this time there was a very real problem of thought. With ideas of evolution and later after the rediscovery of the work of Mendel, it was realised that there must be a physical basis of inheritance. But what was it? How could something so fundamental hide so easily, like the sun behind a cloud. The thinkers of the day were struggling to unify ideas that would give a meaning to inheritance as a concept and a reality.

Darwin knew that to understand evolution it was going to be necessary to have a complete understanding of heredity, not a restating of observed events; and on this he pondered long and hard. What he came up with was basically a resurrection of the old, indeed ancient, theory of pangenesis, but with some effort to accommodate the objections that Aristotle had made against it 2000 years previously. To see where the problems with this idea crop up, it is worth looking at how Darwin set out his plan of inheritance (Fig. 1.5).

Pangenesis requires every cell to produce gemmules or pangenes, each gemmule corresponding to a specific cellular type. These various gemmules are then dispersed from their origins and accumulate in the sex cells. This means that each spermatozoa or ovum will contain gemmules from every single cell type. Upon

Fig. 1.5 Charles Darwin. 'Men of the Day' Vanity Fair magazine 1871

fertilisation these gemmules combine to recreate all the new cell types in the complete organism. As this is a functional description of what was observed to happen masquerading as explanation, without experimental data, it was a matter of faith. Under these circumstances, it is not only possible but also easy to answer the objections as there is neither supporting nor refuting objective data.

Within pangenesis as an idea was a major problem which revolved around the idea that some characters could 'skip' a generation. If a large enough group over sufficient generations is taken, it can be seen that this process of 'skipping' forms a discernable pattern. If taken just as an occasional observation, in one family, for

Fig. 1.6 Jean-Baptiste de
Lamarck by Charles
Thevenin c.1802

example, the picture is far more confusing. So what Darwin suggested was that some gemmules would be inactive in one generation and become active in subsequent ones. In this way he supplied at least a part solution to the Aristotelian objection that a child might more closely resemble its grandparents than parents.

Part of the confusion with Darwinian ideas at this time was the belief that they should be taken to an explanatory conclusion. Unfortunately, postulating gemmules as agents of inheritance required a certain acceptance of acquired characteristics being at least in part heritable. This was broadly in contradiction of the central tenet of evolution, and although it smacks of Lamarckism, it is fundamentally different. Lamarckism was the brainchild of Jean-Baptiste Lamarck, a very clever naturalist who made important contributions to classification and variation but then ruined his own historical reputation. He managed this with his hypothesis to explain evolution. This could be put in a nutshell as characteristics can be attained during an animal life and passed on to further generations. Thus, the giraffe would have got its neck by stretching up to reach food higher on the trees. This requires, by implication, that there is a goal of a long neck which can be aimed for (Fig. 1.6).

Lamarckism had a teleological element, which evolution as expressed by Darwin does not. The idea of a guiding hand is pivotal to Lamarckism but is dismissed as unnecessary in Darwinian evolution. Darwin saw the benefit of random variation, while Lamarck suggested the plant or animal was guided by a will with regard to how it adapted. It is worth remembering that in Darwinian evolution, environment acts upon variation; it does not create the variation in any way whatsoever. This is the difference between the Darwinian idea and Lamarck. Put another way, variation occurs; sometimes the variation fits the conditions well and gives an advantage, but there will be many more variations which don't. Some will probably even result in death of the individual carrying the characteristic. It is pivotal that in evolution the variation is random in occurrence, whereas the variation in Lamarckism is specific to the environmental pressure that requires change for survival.

Such was the state of knowledge when Darwin was alive. It was based on units of heredity that had never been seen and was in some ways simply made up to explain the observed facts. It did indeed help with explaining livestock breeding, but only in a strictly empirical way. What this revolution in thought was creating a change away from teleology towards discrete units of inheritance being responsible for how things were.

Into this arena Mendel started his work on hybridising plants. Until this time plant hybridisation had been undertaken in a broadly haphazard way, well, at least, a slightly haphazard and empirical way. But by the time Mendel published his results, he not only had a clear idea of the process of inheritance, but he also knew which questions he could ask and expect to get an answer from his experiments. The experimental model of choice was the pea, and it was not by chance that he worked with these. He worked with an experimental organism that he could expect to produce quantifiable results from and one that he was familiar with as a plant.

It is said that he had been in contact with Nägeli, a botanist who had suggested he looked at hawkweed for his breeding experiments. This really would not have worked at all. If you do not know hawkweed, it is a sort of tall multiflowered dandelion, a very common meadow flower in the summer. Now think about why Mendel would have failed if he had tried to use it as a plant for his experiments. Not as one would imagine because of the possibility of cross-fertilisation from the plants over the wall in the next door field, it is quite the contrary. Hawkweeds don't cross-fertilise; in fact they do not even self-pollinate. What they effectively do, by a process called apomixis, is clone themselves. They do this not by vegetative reproduction, sending out runners like strawberries or buttercups do, but by seed production. All the dandelions do this. As a result it is quite likely that every dandelion growing in your lawn is a clone of the original one that blew in on the wind. As you can imagine, trying to make any sort of genetic cross with a species like that is doomed to produce results of no value. Actually, since you would be producing endless clones, there would not be any variation to measure anyway.

Although Gregor Mendel could not have known that Hawkweeds had this strange breeding pattern, he did know that peas were clearly much more variable in their growth and expression of size, shape and colour. He was a very astute worker; his choice of plant shows this, even before he started producing results.

The pea that he used was *Pisum sativum* which has seven chromosomes, although of course Mendel did not know this. By coincidence he was investigating seven different characteristics, but these are actually on four different chromosomes. This could have caused him problems, but by the astute investigation of single and paired characters, his results bore out a simple and straightforward method of inheritance. He had also chosen an organism which would not only cross-pollinate but was quite easily self-pollinated as well. The details of his analysis are very well known now; taught in schools at a very junior level, they remain pivotal to an understanding of the principles of genetic inheritance. Genetic inheritance, you note, since other forms of inheritance are known, like titles and money, for example. What he managed to do which was both different and original was to look at single traits. Previously, breeders had bred organisms for an overall effect, larger fruit and more milk, that sort of thing, and it had been carried out in a haphazard and piecemeal manner. As a result of this way of working, a casual observer would naturally assume that traits were not inherited separately but by a process of blending. By looking at single traits for the first time, it was possible to see exactly how inheritance really works.

When the analysis was carried out, Mendel had in his hands results which broadly contradicted all previous explanations of plant breeding. Not only that but he had an insight into the cytological basis of inheritance which was remarkable. He realised that the factors controlling an inherited physical trait, in whatever form it manifested itself, existed as a pair, one from each parent. What makes this stand out as an idea, besides its intrinsic truth, is that he produced it from a mathematical understanding of his data with no physical evidence at all. So his conclusion was that all traits occur in pairs, and they must be divided in half to be passed onto the next generation where they will recombine to form the complete set of genetic material. In this Mendel was predicting the existence of chromosomes without any knowledge of them and also of what we now know as the haploid and diploid nature of cells.

Mendel's results were genuinely a revolution because they did what all science should do—take observations, sit and think to understand and then use that understanding to make predictions of future results which can be tested. Unfortunately the results when they were published were, well, ignored. The article which Mendel published was *Versuche uber Pflanzen-Hybriden* in the *Verhandlungen natur-forschender Verien in Brünn, Abhandlungen* iv pages 3–47 1865. This was the proceedings of the Brünn natural history society and was produced in 1866. Interestingly, Mendel was also down in the journal as being responsible for the meteorological reports for that year. Although it is sometimes said that it was where he published that held up wider acceptance of his work, this is only a small part of the story. More importantly it was radical, it was also mathematics as applied to biology which was unheard of and it was also seen as being very specific in observation, not a general explanation of inheritance.

The article which laid the groundwork of modern genetics was unusual by modern standards but quite normal for its day in that it was long. Nowadays, journals put word and page length restrictions on article length which stifles interpretive explanation of results. This is a development in publishing which took hold in the second half of the twentieth century. Until then it was common-place for journal articles to stretch over many pages, almost to chapter length.

The point at which the true worth of Mendel's paper was recognised was when it was republished in English. A British version was translated by the Royal Horti-cultural Society and published with introductory remarks by William Bateson as Experiments in Plant Hybridization (1901) (Fig. 1.7).

At this stage there was no indication that chromosomes were directly important to inheritance, and there was no proof that nucleic acids were important either. It was a strange time in the nineteenth century for genetics because there were all sorts of investigations going on, with no method of tying the different threads together. So we can see cell biologists and biochemists and geneticists all working at different threads of the same story.

Fig. 1.7 William Bateson

Fig. 1.8 Albrecht Kossel in 1911

One of these steps forward was made when nucleic acid became of interest to scientists. So it was in 1869 that a Swiss chemist, Friedrich Miescher, analysed cell extracts and demonstrated that they were made up of proteins and nucleic acids (Miescher 1871). It was not long after this that Albrecht Kossel analysed nucleic acids in more detail to show the basic components, which he named adenine, guanine, cytosine, thymine and uracil (Kossel 1881)(Fig. 1.8).

These results were the products of chemical analysis, with no thought regarding the way this information sat within the framework of the cell. As much as anything, the cell as a structure was still a more or less closed book. It took the work of a very skilled observer, a zoologist, Walther Flemming (1843–1905), to start the ball rolling in a meaningful way regarding cell and nuclear division in animal cells.

Although he may not have been the first to observe mitosis, he was certainly the person who brought attention to chromosomes as important parts of the cell. It was Flemming who coined the word mitosis, from the Greek word for thread, referring to the observation of chromosomes, but it was not Flemming who came up with 'chromosome'. It was in 1888 that Wilhelm Waldeyer-Hartz introduced the Greek for 'coloured body' as a word, by creating a composite noun, and so coined chromosome. Now they were recognised as components of cells, important ones worthy of investigation in their own right.

Flemming had noted the intensity of colour at various stages of nuclear activity and had described this as chromatin. A testament to his skill as an observer can be found in his description of prophase when he said that the chromatin was double stranded. This is extraordinarily difficult to see, even for an experienced observer with modern equipment. It should be noted that although Flemming was able to observe and describe much of the cell cycle (Flemming 1878, 1882), terms such as prophase, metaphase and anaphase were not used until the noted botanist Eduard Strasburger coined them in 1884. Strasburger showed convincingly that chromosomes behaved in a manner identical in plants as animals (Strasburger 1876, 1884, 1894).

It is impossible to imagine that these men of science would not have realised, or at least speculated, on the significance of chromosomes to the process of inheritance. Indeed, it would be true to say that Flemming and Strasburger sparked a considerable debate over the question of whether chromosome numbers were constant. This was a question not just of constancy between species, which was expected, tissues and organs within an individual. This was not so easily decided because although individuals within a species are broadly similar, cells are not. Organs are not only different in exterior appearance, but microscopes were starting to show that cells in different organs are so far removed from each other that they looked as they did not belong in the same organism. But then it was noted that amongst mammals, for example, the gross morphology of a kidney was always broadly the same. Even more remarkably, the microscopic structure was the same as well. So if the organs were the same, was the chromosome number the same between organs rather than species?

There was an acknowledgement that chromosomes were important—they were always there when cells were looked at. What was not known was why they were important and what function they fulfilled. Flemming had demonstrated to his own satisfaction that the number of chromosomes in salamander epidermis was constant at twenty-four. In this he was lucky in choosing a subject with large epidermal cells. In the same way, Strasburger had been both lucky and prescient in working on hair cells of the house plant *Tradescantia*. The questions regarding the constancy of chromosomes were not entirely sorted out by these workers because other workers reported different numbers of chromosome in different cells from the same species. This in itself is not surprising; visualising chromosomes is not easy. It takes a great deal of expertise and time to be able to routinely produce a reliable product that can be analysed with confidence. If it was easy the human chromosome number would have been worked out much sooner than it was.

Almost contemporary with Eduard Strasburger was another Eduard, this time Eduard Van Beneden. Van Beneden was also lucky in his choice of experimental organism. It is worth considering his choice of laboratory animal because it sounds so implausible. He studied the chromosomes of an intestinal nematode worm of the horse, *Ascaris equorum*. *Ascaris* is the genus of the worm; of course, the horse is *Equus*. This choice by Van Beneden was particularly lucky because in somatic tissues the parasite has four chromosomes which are quite large. In this way he demonstrated the consistency of chromosome number. But better than that, he showed that in sex cells, that is, ova and spermatozoa, the number was half, two. This is what would be expected if the chromosomes had to be halved in number so that on recombination at fertilisation, the full number is restored. This, then, is a physical demonstration of Mendel's idea on inheritance. This change in chromosome number is from a single complement, *n*, called haploid, to a normal, or diploid complement, *2n*. In the case of *Ascaris*, the normal complement is 4, made up of two pairs of homologous chromosomes. They are referred to as homologous because they carry the same genes but not necessarily coding for identical versions of them (Beneden 1868).

So although we can see that there was now both a physical basis of inheritance and an experimental explanation of observed character inheritance, the two would not be reconciled for some time. This reconciliation of scientific ideas would finally take place in the early years of the twentieth century when three scientists, de Vries, Correns and Tschermak von Seysenegg independently rediscovered the work of Mendel. This happened because they were conducting a literature search as background to their own researches. It was William Bateson who pursued the ideas of Mendel most vigorously, defending it against the disbelief that accompanied the results. It was to most breeders not possible to give up long-held ideas of blended characters and immutable species.

What we have to realise here is that in many ways the doubters were right. Single-gene characters are really not very common in the panoply of inherited traits. A simple example will suffice to demonstrate this. If Mendel had been looking at hair colour in people, what would he have seen? A range of tones from albino white to the darkest black but, in between these extremes of range, any shade of blonde, brown and red that can ever be described. Inheritance of hair colour is not so straightforward as we might imagine. It becomes more strange when we consider that many shades of hair change with age, not just going grey, but many light shades darken with age. Simple monogenic inheritance of characteristics is rare, but all genes are inherited in that manner; the expression of them is what makes genetics so intriguing. Not only were there examples of continuous variation which caused confusion, but in 1905 William Bateson demonstrated that some characters do not segregate independently. We know why this happens now; it is simply that the genes are linked, and put another way they are on the same chromosome. In 1900 to suggest such a thing would have appeared to some as another case of fitting the explanation to the observations.

References

Bateson W (1901) Experiments in plant hybridisation. J R Hortic Soc 26:1–32

Beneden Év (1868) Recherches sur la composition et la signification de l'œuf. This article is in French and is a transcript of a presentation at the Acadamie Royale de Belgique in December 1868

Darwin C (1859) The origin of species by means of natural selection, or the preservation of favoured races in the struggle for life. John Murray, London

Darwin C, Wallace AR (1858) On the tendency of species to form varieties; and on the perpetuation of varieties and species by natural means of selection. Zool J Linn Soc 3:46–50

Flemming W (1878) Zur Kenntniss der Zelle und ihrer Theilungs-Erscheinungen. Schriften des Naturwissenschaftlichen Vereins für Schleswig-Holstein 3:23–27, This paper is in German

Flemming W (1882) Zellsubstanz, Kern und Zelltheilung. Verlag von F.C. Vogel, Leipzig

Hooke R (1665) Micrographia: or some physiological descriptions of minute bodies made by magnifying glasses. J Martyn and J Allestry, printers to The Royal Society, London

Kossel A (1881) Untersuchungen über die Nukleine und ihre Spaltungsprodukte. Strassburg, Karl J. Trübner, This book is in German

Mendel G (1866) Versuche über Pflanzen-Hybriden. Verh Naturforsch Ver Brünn 4:3–47 (in English in 1901, J R Hortic Soc 26:1–32)

Miescher F (1871) Ueber die chemische Zusammensetzung der Eiterzellen. Hoppe-Seyler's med Chem Unters 4:441–446

Schwann T (1839) Microscopic investigations on the accordance in the structure and growth of plants and animals. Berlin (English translation by the Sydenham Society, 1847)

Strasburger E (1876) Über Zellbildung und Zelltheilung [On cell formation and cell division]. Nabu Press. In German, Reprinted 2010

Strasburger E (1884) Das botanische Practicum. First published in German, 1884

Strasburger E (1894) Ueber periodische Reduktion der Chromosomenzahl im Entwicklungsgang der Organismen. Biologische Centralblatt 14, This paper is in German

Waldeyer-Harz H (1888) Über Karyokinese und ihre Beziehungen zu den Befruchtungs-vorgängen. Archiv für mikroskopische Anatomie und Entwicklungsmechanik 32:1–122, This paper is in German

Microscopes and Stains: The Rise of Technology

Before any coherent ideas about cells and their contents were going to be possible, they needed to be seen. In the twenty-first century, we are commonly assailed with images that have been altered by computer with artificial colours and enhanced contrast. I have made use of these techniques myself and they are very useful too. These images are of such staggering clarity that it is sometimes difficult to be sure what is real and what is not. It is often not fully understood or remembered that to see details of a particular cellular structure, it is usually only possible if the cell is killed and stained in various ways. Once that is done, viewing another component of the same cell becomes very difficult. Without the technology, both physical and chemical, to see that a cell is far more than just a bag of liquid, it was commonplace to imagine that the tissues and organs of the body were complete in themselves and irreducible. That organs are more than just active sacks only became apparent with the rise of some truly magnificent engineering.

Some cells are individually large enough to be seen with the naked eye, but these are very few; a bird's egg is an example of such a thing. It was impossible without some aid of observation to conceive that every living thing was made up of cells. In fact even the word cell was not applied to biological material until Robert Hooke first used it in *Micrographia* (1665). He created its biological meaning from the original Latin root for a small room. The tool which allowed this leap of understanding in biology was the microscope, of which Hooke was not only a pioneer but a supreme exponent and observer.

Microscopes had been around in various forms for many centuries before cell biology emerged as a science, but unlike today, if you wanted a microscope, it was incumbent on you to build it yourself. Alternatively you could always pay someone else to make one for you, which did tend to elevate the early microscopes to the level of expensive toys of the wealthy.

One of the early pioneers of microscopy was Antoni Van Leeuwenhoek. Born in 1632 at Delft in the Netherlands, it was there that he pioneered microscopy. At this time the idea of a career in science was virtually unknown, and many of the great scientific thinkers of the day were either gentlemen of independent means or

© Springer International Publishing Switzerland 2016
W.J. Wall, *The Search for Human Chromosomes*,
DOI 10.1007/978-3-319-26336-6_2

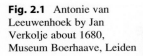

Fig. 2.1 Antonie van
Leeuwenhoek by Jan
Verkolje about 1680,
Museum Boerhaave, Leiden

variously employed at other tasks to make a living. It was for this reason that the
Leeuwenhoek Haberdashers in Hippolytusbuurt Street, Delft, funded his very
serious hobby of microscopy. It should not be imagined that Leeuwenhoek's
interest in lenses arose from nowhere; checking fine detail of cloth was an important
part of a merchant's work and for this, magnifying lenses were extensively used
(Fig. 2.1).

After about 1660 he began making simple microscopes, that is, ones with single
lenses, rather than compound microscopes which have multiple lenses. He was not
the first microscopist, but his skill was outstanding and the results were difficult to
reproduce by those less practised at observing and seeing what they are looking at
(Dobell 1932).

Working alone, though corresponding with many learned men of the day, the
work of Leeuwenhoek was for a long time denigrated. His microscopes were being
regarded as little more than curious toys. There is something else about discerning
details within cells which should not be forgotten, and that is consistency. A cell is a
dynamic and working piece of biology; in the seventeenth century when nobody
had seen the internal structure of a cell, you would be forgiven for not believing
your eyes. This attitude regarding simple microscope as a toy might well be because
until the nineteenth century, the more complicated and ornate compound
microscopes were seen as playthings of the rich, expensive and often poorly used
instruments. Even though compound microscopes had been available before 1650,

they were expensive but inferior devices. Consequently a simple microscope, a single lens in a simple holder, could not possibly be worthy of serious consideration. What had been forgotten was that the observer is at least as important as the equipment, and Leeuwenhoek was a meticulous and skilled observer using his single-lens microscopes.

Another extraordinary observer and microscopist was Robert Hooke (1635–1703). He was born at Freshwater on the Isle of Wight and the son of the local minister. Interestingly, the birthplace of Hooke is now commemorated by a stone in the grounds of the Portsea Island Co-op in Freshwater. There are no contemporary images of Hooke; all the pictures we have of this luminary are later productions, often based on written descriptions of him.

Although Hooke favoured the compound microscope over the simple microscope, these were of relatively low power, and he admitted that for high-power work, it was necessary to use a single-lens microscope. This was probably because with little or no theoretical basis for manufacturing lenses, they would tend to cause colour aberrations and spherical distortions in the image, so although putting the two lenses together would increase the magnification, it would simultaneously magnify the distortions by a corresponding amount.

With the advent of the nineteenth century, attitudes were to change regarding microscopy; this was the century when microscope optics became a scientific discipline and with it widespread and repeatable observations. At this time the best and most consistently reliable telescopes were undoubtedly made in London by the firm of Dollond and Sons. These telescopes were still constructed using empirical techniques, entirely dependent on the person putting the finished product together for the quality of the instrument (Fig 2.2).

This company had its origins with Peter Dollond (1730–1820) and his father John Dollond (1706–1761). It is surmised that the original spelling of the name was d'Hollande, later anglicised for ease of use and assimilation into the area of east London where many such as themselves had been chased by a religiously intolerant French nobility. As a family of silk weavers, John continued the tradition while developing a keen interest in optics. It was his son Peter who, having inherited his

Fig. 2.2 Potrait of Peter
Dollond by John Hoppne

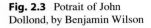

Fig. 2.3 Potrait of John
Dollond, by Benjamin Wilson

father's skill and interest in optical instruments, opened an optician centre near
Strand, London, in 1750. At this point his father joined him in the opticians venture.
Although highly skilled at manufacturing lenses and telescopes, the company as a
whole had very little theoretical knowledge to back up what they were doing. The
company evolved to Dollond and Aitchison which became a significant UK retail
optician business until it was subsumed by a larger company in 2009 (Fig. 2.3).

Further commercial development of the company, indeed all optical companies,
was severely hampered from an unexpected quarter. In 1745 the parliament
introduced the glass excise tax, sometimes erroneously mixed up with window
tax. It was in fact a direct tax on glass and a very complicated one at that. It was the
complexity and interpretation of the rules which caused such severe problems for
British glass manufacturers. Broadly speaking every glass melt was taxed, so
companies such as Dollond, who regularly remelted the glass of failed lenses,
found themselves paying tax on the same glass several times over. One commercial
result of this tax was that lenses were imported from Switzerland and France;
although often known to be inferior to the local product, the difference in price
far outweighed this.

When the glass tax was finally repealed in 1845, glass manufacturers were no
longer hampered in their development of new products and new types of glass as
they were no longer taxed on their failures. Glass could be remelted until it finally
made it into a product, without additional excise costs being involved. In fact it
made such a difference to the glass makers that soon afterwards the then president
of the Royal Microscopical Society declared in a speech that the finest glass for
lenses was made by Chance Brothers and Co., Smethwick, in Birmingham. This
company is still in existence making microscope slides and cover slips, but in 1850
they received an order of quite a different magnitude, glass for the Crystal Palace to
be built for the great exhibition of 1851. Chance Brothers produced the 400 tons of
sheet glass that was required to glaze it. Interestingly Chance Brothers were used to
producing large quantities of glass, such as the lenses for lighthouses (Fig. 2.4).

Fig. 2.4 Heceta Head Lighthouse in Oregon, USA. The Chance Brothers Fresnel lens, built in the early 1890s, is still in operation at this historic light house

Microscopes for biology and biologists were becoming essential equipment as designs improved and the quality of the products became more reliable. They were moving away from being curiosities into the mainstream of scientific research, fuelled by an ever-increasing confidence in the ability of man to wrestle the deepest truths from nature. Even so, in the first half of the nineteenth century, it was still thought that microscope optics were so complicated that they would defy mathematical analysis. Therefore, it was still the skill of the manufacturer that determined the quality of the product, and every microscope was as individual as the user.

As demand for good quality microscopes increased, so, too, did the number of companies making and selling them. The nineteenth century saw a considerable rise in numbers of both manufacturers and retailers. In 1800 there were about 40 such companies in the UK, 75 % of which were based in London. By 1810 this had risen to 50, and in 1840 there were 60 companies dealing in microscopes, of which 10 were manufacturers and the rest retailers. Not only was the trade based in and around London, but it was also more highly developed in Great Britain than any other country. In 1865 Great Britain had 16 manufacturers, France 5, Germany 7 and Austria and Italy one each. This was an extraordinary economic trend because nobody could have foreseen the demand for precision optical instruments of this type when many of these companies were just starting.

Similarly the trend in size of companies making this type of equipment was to get bigger as they encouraged and supplied demand. Leitz, for example, had 10 employees and made 31 microscopes in 1851, while by 1900 this had increased to 400 employees and 4000 microscopes.

Instrumental in developments of microscopy was a group of 17 individuals who met informally at the house of Mr. Edwin Quekett at 50 Wellclose Square, London.

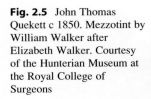

Fig. 2.5 John Thomas
Quekett c 1850. Mezzotint by
William Walker after
Elizabeth Walker. Courtesy
of the Hunterian Museum at
the Royal College of
Surgeons

John Quekett

Although Wellclose Square is still there, sadly number 50 is not, having been demolished many years ago. The significant event for microscopy that took place at 50 Wellclose Square on 3 September 1839 was that it was decided to form a microscopy society, the first of its kind in the world. This eventually went on to gain a Royal Charter from Queen Victoria and become the Royal Microscopy Society.

Also important at the instigation of the new society dedicated to microscopy were three brothers, the Queketts. Although of diverse occupations, they all shared a love of microscopes and microscopy. These three brothers were Edwin, William and John (Fig. 2.5).

John Quekett was secretary to the Royal Microscopical Society for 20 years, taking up the position in 1841. Of the three Quekett brothers, it is probably John who is best known for his microscopy. When John was only 16 years old, he had constructed a microscope with which he gave lectures. When he left school, he was first apprenticed to a local surgeon and then later to his older brother Edwin, who was already in a responsible position.

Once established at the Royal College of Surgeons, John Quekett started to keep a diary which is very instructive regarding the development of the nineteenth-

Fig. 2.6 Robert Brown

century microscopy. In 1841 he said that he went to a 'microscope party,' with many notable microscopists present, such as Powell, a manufacturer; Lister, the practical and theoretical microscopist; and two botanists. The first was Conrad Loddiges who had a plant nursery in Hackney, east London, where he was one of the first people to offer orchid plants for sale to the public. The other botanist present was Robert Brown who observed what we know as Brownian motion (Brown 1828; Fig. 2.6).

One of the most extraordinary investigations that John became involved in took place between 1848 and 1849. During this period he undertook an investigation of particles of skin and hair taken from various ancient church doors. He managed, by microscopy, to show that this material was of human origin. The importance of this was that it was reported that during previous centuries captured pirates were flayed and their skin attached to certain church doors. Although Quekett concluded that the material was human in origin, he probably arrived at this result in conjunction with the circumstantial evidence of finding material which fitted in with the belief that human skin had been attached to church doors in the first place.

The entire enterprise of revealing the nature of chromosomes is indissolubly linked to the development of the microscope. These three brothers were instrumental in setting up the Royal Microscopical Society, and the society was important in setting standards that have allowed all professional microscopists to work with ever greater accuracy and precision. The problem of standards is easy to state, but it took an umbrella organisation in the form of the Royal Microscopical Society to solve them. During the major part of the nineteenth century, every manufacturer of microscopes had ploughed their own furrow when it came to microscope slide size and threads for attaching lenses to microscopes. Consequently, when microscopes changed from being sophisticated toys for the wealthy amateur to essential tools in the rapidly developing biological sciences, most microscopists were faced with some major problems, long before they could possibly begin the observation of chromosomes.

These difficulties lay with it being virtually impossible for equipment from one manufacturer to be used with the equipment of another. This extended from being unable to swap eyepieces and objective lenses between instruments to being unable to use microscope slides from one instrument to another. It was this latter problem which was the first to be approached by the Royal Microscopical Society. What was decided was that either of the two sizes should be used. These were 3 inches × 1 inch or 3 inches × 1½ inches. Interestingly, it was the manufacturers that finally drove the 3 × 1½ inch microscope slide to extinction. Even now boxes of micro-scope slides are labelled as 76 mm × 26 mm and 3 inches × 1 inch. This was a major step forward for microscopists because they could now exchange slides and know that they would be useable, but better was yet to come.

Over the following years, it became apparent that even such apparently mundane subjects as the cut of a thread to join a lens to a microscope needed to be standardised. This was strangely important because while some microscope manufacturers made exceptionally useful and robust stands, others made excellent objectives and others eyepieces. So to create an excellent microscope, those that could not afford custom-made equipment put together their own versions of creat-ing a composite device. This, of course, would need various linking parts, all of which required their own threads and methods of attachment.

What happened over the following years was that the Royal Microscopical Society commissioned Joseph Whitworth to make gauges for the threads used for various parts of microscopes. For some time the Royal Microscopical Society was effectively the keeper of these standards, although now it is the prerogative of the International Standards Organisation (ISO).

But while British microscope manufacturers of the nineteenth century took on board these standards, it was the new commercial power of Carl Zeiss and his acceptance of them that made microscopy an internationally standardised and important science. This helped to move microscopy from simply reporting observations which were difficult to reproduce to a sound and repeatable scientific method.

There were many nineteenth-century microscope manufacturers that were to be significant in providing microscopes which would be used to investigate biology in general and chromosomes in particular. Investigations of chromosomes and their numbers would take place in both plants and animals, not just humans. One of the most revolutionary of the microscope manufacturers, both scientifically and com-mercially, was Carl Zeiss. This company was originally set up by Carl Zeiss, but it was definitely not just his work which resulted in the groundbreaking microscopes which came out of Jena in Germany and pushed forward the understanding of optics so that now any student user of a microscope should be able to visualise objects at the very limit of resolution of the instrument (Fig. 2.7).

Carl Zeiss was born in Weimar on 11 September 1816, the son of a toy shop proprietor. It is difficult for us to appreciate the status of instrument makers in the nineteenth century, but these were highly regarded and highly skilled individuals; such a man was Friedrich Korner. So it was that as an apprentice to Friedrich Korner, at the University of Thumingen in Jena, Carl Zeiss learnt many skills and went on to work with instrument makers in Stuttgart and Vienna (Auerbach 1925).

Times were changing and so were the fortunes of the ambitious Carl. In 1846 Zeiss founded a workshop in Jena as instrument maker, and it was here that a biologist at the university influenced the direction in which Zeiss would go. Jacob

Fig. 2.7 Carl Zeiss, 1885. Carl Zeiss Archive

Fig. 2.8 (**a**) Matthias Jacob Schleiden c.1855 by Carl Schenk (**b**). Theodore Schwann

Schleiden, already an influential biologist responsible with Theodor Schwann for the cell theory, persuaded Zeiss to concentrate on optics. There is little doubt as to the influence that Schleiden had, both with the authorities and with Zeiss, as it was in part due to him that Carl was allowed to open his first workshop, helped by a loan from his brother (Fig. 2.8 a and b).

When he first started his workshop, he was on his own, not only making instruments but also repairing them. By 1847 he had introduced a single-lens microscope for sale. Being a single lens probably demonstrated his awareness of magnified aberrations in compound instruments as well as the essential portability of single-lens devices. They would, of course, also be both easier and cheaper to manufacture.

Zeiss was constantly trying to improve his lenses and microscopes but was hampered by a very simple thing: his lenses were constructed using skilled craftsmen by trial and error. He was convinced that this was not the best way, after all telescopes had been constructed using scientific principles for many years, so microscopes, he reasoned, should also yield to principles of geometric optics. Even so, the well-known and experienced mathematicians of the day were unable to help in the complicated area of microscope optics.

Things were about to change, however, as there was at this time a young 26-year-old lecturer at the university in Jena called Ernst Abbe. When Zeiss asked Abbe if it was ever going to be possible to mathematically design a compound microscope lens, Abbe took up the challenge, even though he had not until then been involved in optics.

It was not as though geometric optics was an unknown area of investigation; the simpler construction of telescopes had been proceeding well using mathematical design. But microscopes were a completely different field, with different requirements of magnification and focusing. Early attempts were not always as successful as would have been hoped. Once designed the lens would be made to a set specification, even if the result was only fit to be used as a paper weight!

By 1872 Abbe had solved many of the earlier problems associated with his lenses. So important was this to the following microscope manufacturers that his mathematical formulation relating to resolving distance of a lens to both the wavelength of light and refractive index is inscribed on his memorial in Jena (Hammond 1996; Fig. 2.9).

Fig. 2.9 Ernst Abbe, centre; Otto Schott, right; and Paul Rudolph, designer of planar lens, with his chainless bicycle, outside Abbe's house. 1898, Carl Zeiss Archive

One of the primary things that Abbe realised is that the glass from which a lens was made affected its performance; this may seem self-evident now, but then it was generally felt that within broad limits, glass was glass. When he subsequently visited an exhibition of scientific equipment in 1876, his review of the event resulted in a chemist, Otto Schott, getting in touch with him. Schott was a member of a family of glass makers, but more than that, he was interested in developing new forms of glass for application to optics rather than simply window glass. It is even said that Abbe could specify glass of a particular refractive index and Schott was able to make it. Whether this is true or not, the result of these collaborations was that the Zeiss company produced benchmark microscopes that outperformed others available.

This is not to say that other microscopes were not potentially as good; indeed in 1884 E. M. Nelson had attained the ultimate resolution of standard transmission light microscopes using a Powell and Lealand No 1 'Jug-Handle' (Bradbury 1998). The difference between this and what Zeiss was offering was that with a Zeiss 'off-the-shelf' microscope, you were able to render consistently good results. Such was the commercial pressure of Zeiss that by 1910 Powell and Lealand had more or less given up the production of microscopes.

So by the end of the nineteenth century, microscopes had the potential to see everything that a light microscope can see. So why was it not possible to count the number of chromosomes in a cell? Well, there is more to counting chromosomes than just magnifying them. The preparation of the sample is paramount. This problem of technique was to dog the whole science of cytogenetics for the next hundred years. As a start, the production of an adequate specimen is of vital importance. A first step in this direction was the production of ancillary equipment to manipulate the sample and the stains to make them visible.

Microscopy has never been simply a case of putting a sample under a microscope and looking at it. It would also be a mistake to imagine that the most important job of a microscope is magnification. A microscope has three important attributes: magnification, resolution and contrast. Of these three, contrast is out of the hands of the microscope manufacturers, although it is probably the most important aspect of a sample. It is magnification and resolution that interested the earliest makers of microscopes. However, without decent methods of preparation any microscope would be useless.

Until the middle of the nineteenth century, production of specimens for microscopy was based on a do-it-yourself system. Indeed, until recently, methods of producing single-cell thick samples by hand for microscopy from botanical specimens were still being routinely taught in colleges. The process is rather dependent upon the skill of the operator, but considering the equipment the result can be really quite good. The equipment in this case is a cut-throat razor, not a barber's razor, hollow ground on both sides, but one with one curved side and one flat side.

Into this world of developing microscopes and undeveloped specimen preparation came Horace Darwin, born in 1851 and later to become Sir Horace. He was the son of Charles Darwin and really something of a maverick within the Darwin family. Graduating from Cambridge in 1874 having studied under a Scottish Physicist James Clerk Maxwell, Horace had made friends with another student, Albert George Dew-Smith.

Horace Darwin returned to Cambridge in 1877 where in 1881 he and Dew-Smith founded Cambridge Scientific Instruments. This was to be a very significant event for biologists. The device which was to prove so important for biologists was the Cambridge Scientific Instruments microtome. Microtomes were to replace the razor as a routine method of cutting sections, the aim being to produce very thin sections consistently and reliably, which as can be imagined was very difficult by hand. Microtomes were already available but were unreliable. It was thought that by sectioning through cells, it should be possible to reconstruct the entire nucleus and thereby count the chromosomes. This would require an extraordinary piece of equipment which could be relied upon to produce serial sections of even thickness. Nobody had made serial sections at this time and for another the sections were all too thick. Even when the section is thin enough to let light through, the internal structures and overlapping cell membranes from adjacent cells make the resulting image virtually impossible to interpret.

Realising that they could make a better instrument than the one they sold as agents, Cambridge Scientific Instruments produced the first automatically advancing microtome. This is the one that could cut a section and then increment the specimen to cut another of identical thickness. The instrument was deceptively simple yet required high-quality engineering allowing serial sections to be cut down to a thickness of only 1/4000th of an inch (0.0006 mm). So robust were these instruments that it was not unusual to find them working reliably for in excess of 50 years.

Here, for the first time, was an instrument that could produce specimens for microscopy which would complement the resolution of the microscopes that were by then routinely available to the increasing numbers of biologists. It is interesting to note that although the terms botany and zoology had been in common use for many years, it was not until Lamarck and Trevirons simultaneously coined the word biology in 1802 that a unifying term was available for living things.

Using the newly devised reliable microtome and high-resolution microscopes, the hunt was on for details within the cell, the nucleus and eventually the number of chromosomes in a human cell. Much remarkable work was carried out in this field by many remarkable men, but ultimately the answer would not be found with any certainty using specimens generated by a microtome, even though many workers came close to the answer in this way.

It is axiomatic that contrast in an image, any image, is important. Extreme lack of contrast can be white on white or black on black; in these examples, the image tells us nothing about the relationships between the two white objects or black objects; they simply cannot be seen against the background. This becomes a very practical problem with biological material; when whole organisms are magnified, they sometimes contain their own colours and structures which give contrast. But when early microscopists started to take sections of samples, it soon became apparent that biological material has a very low intrinsic contrast. Without contrast no details could be made out and it did not matter what the resolving power of the microscope was, or the magnification, no detail would be visible. With the human eye capable, at best, of determining a contrast difference of 2 % and more normally somewhere between 5 and 20 % depending on the size of the object being looked at and the lighting conditions, the specimen was essentially invisible against its background.

What was needed was a method of increasing the contrast in the specimen. There are really only two ways of doing this, both of which are still used. One is to manipulate the light path in the microscope to create an image from optical interference and the other is to stain the sample differentially. The first of these two was not a practical proposition during the nineteenth century, but the second one was.

From very early times, microscopists had utilised natural dyes, some of quite surprising origins. During the nineteenth century, carmine from cochineal insects, damson juice and even port wine were used as stains. Most of these would fade under strong light. Before 1850 only about a dozen useable dyes were available to microscopists. These were mainly used by cloth dyers who extended the range of colours available to them by skilful mixing, which was not generally practicable when it came to specimens for microscopy.

There are two things which histological dyes require, the first is to be optically dense and the other is to differentially stain. In many stains these can be traded off against each other to produce remarkable effects. The best stains for histology are those that result in substantial coloured precipitation in specific tissues or specific chemical binding with tissue types (Turner 1989).

It should not be imagined that histochemistry was unknown before the middle of the nineteenth century. Even so histochemistry and histology remained hampered by the lack of strong dyes in a range wide enough for two or more to be used on the same specimen. One of the problems with naturally occurring dyes which were sometimes used in histology was that with very thin samples, or very small organelles such as chromosomes, there was simply not enough dye present to absorb enough light to induce adequate contrast in the sample; in physical terms the molar absorption of the dye was too low.

A great change took place in 1856 when the 18-year-old student, William Henry Perkin, accidentally produced a new dye. The importance of this, to the cloth manufacturers and the histologists of the day, was that this was the very first artificial dye. Perkin had always been attracted to chemistry and he was attempting to synthesise quinine as an antimalarial.

The way that Perkin ended up kick-starting the dye industry was that he was trying to make quinine from allyl toluidine. Such was his enthusiasm that he set up an admittedly basic laboratory at home so that he could continue his research in the evening and during vacations. The logic of his attempt to make quinine was sound, although with our current knowledge of chemistry, we can say that it would not have worked. Empirically the reaction he was attempting can be written as:

Fig. 2.10 William Henry
Perkin

$$2(C_{10}H_{13}N) + 3\ O \rightarrow C_{20}H_{24}N_2O_2 + H_2O$$

The reaction that Perkin attempted used sulphuric acid and potassium dichromate, but the result was not quinine; it was a reddish brown precipitate, so Perkin tried the same thing again only starting off with aniline. This time the precipitate was a black sludge, part of which was soluble in water, giving a deep purple solution. This was mauvine, the first of a long line of artificial aniline dyes. Out of curiosity Perkin tried to dye a piece of silk, which resulted in a brilliant mauve which would not wash out or easily fade. He started manufacturing mauve in a shed at the bottom of his garden, literally, but quickly outgrew the facility. What he needed was a factory, so with the financial assistance of his father and the help of his brother, a factory was built and in 1857 it started to make 'aniline purple' (Fig. 2.10).

It was a very versatile dye; even stamps were printed using it. Although primarily used for dying cloth, it soon became obvious that it was also suitable for staining biological specimens. Although the family of aniline dyes were used for textiles, the idea rapidly caught on that there was a group of dense and highly coloured dyes that may be suitable for biological work, and so they became a widespread investigative tool (Garfield 2000).

Later developments in dye chemistry meant that the range of dyes became enormous. In the nineteenth century, many dyes were produced which are still used today. Examples include methylene blue, methylene green, malachite green and Congo red. All of these dyes were originally destined for the textile industry, but they were rapidly spotted for their potential as microscopical stains.

In 1892 the chemical company, Merck, produced a list containing 51 artificial phenol and aniline dyes. As the proliferation of dyes and stains carried on, confusion set in as to the exact nature of the stains and dyes being used by histologists.

Sadly, it is still often the case that histologists are unaware of what they are doing chemically: they simply follow recipes. Such is the reliability of the stains and techniques that they use wherein the results are still dependable.

At the end of the twentieth century, there were approximately 3500 dyes with a commercial value, only a few of which are used in microscopy, the bulk being purely textile dyes. When the dye industry really took off at the end of the nineteenth century, every dye manufacturer used their own naming standards. For the textile industry, this was not a particular problem as they were only interested in the colour, rather than the chemistry. For histologists, however, it was vital to know what the chemical composition of the stain was. Because the chemical industry closely guarded any chemical secrets, especially with new and exotic colours, this became more and more difficult for histologist to find out.

Against this background of confused colour mixes, an attempt to ease the problems for all concerned was made. In 1922 the Society of Dyers and Colourists in Bradford, England, produced the first edition of the Colour Index. It is now a joint production between the Society of Dyers and Colourists and the American Association of Textile Chemists and Colourists. What the Colour Index does is assign a five-digit number to every stain or dye, based purely on the chemical structure. This means that if you were to buy a stain for microscopy, it would not matter what it was called on the label; if the Colour Index number was there, you could know for sure what the chemical composition was. Methylene blue, for example, has the number 52,015 but is sold variously as Solvent Blue 8, Swiss Blue and even tetramethylthionine.

With the technology of microscopy and specimen preparation in place, the provision of high-quality, densely coloured stains for microscopy made the emerging science of cell biology ripe for expansion. For the first time, it was possible to clearly make out subcellular organelles, specifically stained to increase their contrast. Chromosomes were visible and questions were being asked about their significance and function within cells. It was quickly realised that they seemed to be present in every cell that was studied, so clearly they were important, but as what? And since they were obviously separate entities, was the number significant and stable throughout an organism or even across species?

References

Auerbach F (1925) Das Zeisswerk und die Carl-Zeiss-Stiftung in Jena. Gustav Fischer Verlag, Jena. Reprinted in English as The Zeiss Works and the Carl-Zeiss Stiftung in Jena; Their Scientific, Technical and Sociological Development and Importance. Forgotten Books 2012

Bradbury S (1998) E.M. Nelson (1851–1938) a dedicated microscopist. Proc RMS 33:15–17

Brown R (1828) A brief account of microscopical observations made on the particles contained in the pollen of plants. Lond Edinb Philos Mag J Sci 4:161–173

Dobell C (1932) Antony van Leeuwenhoek and his "Little Animals": being some account of the father of protozoology and bacteriology and his multifarious discoveries in these disciplines. Harcourt, Brace and Company, New York. Reprinted 1960 (Dover Publications)

Garfield S (2000) Mauve. Faber and Faber, London

Hammond C (1996) 150 years innovation in optics 1846–1996. Meeting to commemorate the 150th anniversary of the foundation of the Carl Zeiss workshops in Jena. Proc RMS 31:279–299

Hooke R (1665) Micrographia: or some physiological descriptions of minute bodies made by magnifying glasses. Royal Society, London, J Martyn and J Allestry, printers to The Royal Society

Turner GLE (1989) The great age of the microscope: the collection of the Royal Microscopical Society through 150 years. CRC Press

Mendel and Genetics

3

It was known that chromosomes were important to cells; they always seemed to be there, but it was not at first realised why they were important. It is well known that Mendel formulated the laws of simple inheritance, but what is frequently forgotten is that he had no knowledge of chromosomes or their method of segregation. What he did have, however, was the insight to recognise that some characteristics are inherited in a discrete manner; there are no intermediate forms. This was pivotal for his research as trying to decipher blended characteristics and the underlying mode of inheritance is a very difficult task. There is little doubt that his choice of characters to deal with was not just fortuitous but came out of long observation. He joined the Augustinian order in 1843 at the age of 21 and was ordained in 1847, but his plant experiments did not begin in earnest for some years. In 1851 he went to Vienna for 2 years to study science, and it was on his return that he took up the mantle of plant breeder, having already failed twice to pass the teachers' qualifying examinations (Fig. 3.1).

In his original paper, he looked at seven different characteristics of pea plants, *Pisum sativum*, each one being clearly defined so that there should be no intermediate forms; they are either one form or the other. Even the tall and dwarf characters are separated by plants being either 50 cm tall or 2 m tall, a distinct difference once the plants mature. This would appear to be a sufficient distinction to enable a good statistical collection of data to be made, but if we look at the list of characters, we can recognise some areas of difficulty (Mendel 1866).

The seven original characters were:

1. Difference in form of the ripe seed. These were described as round or roundish with any depressions being shallow, or they are irregular and deeply wrinkled. He showed the round form to be dominant over the wrinkled seed.
2. Difference in colour of the endosperm. Actually, he was referring to the cotyledons as the endosperm of dicotyledonous plants which are frequently triploid. The two colours were either pale yellow, bright yellow/orange or a bright green. In this case it is the yellow colour which is dominant.

© Springer International Publishing Switzerland 2016
W.J. Wall, *The Search for Human Chromosomes*,
DOI 10.1007/978-3-319-26336-6_3

Fig. 3.1 Gregor Mendel,
1822–1884. Wellcome
Library, London

3. Difference in colour of the seed coat. Here the colours are described as either white or grey, grey brown or leather brown. Interestingly, he associated the colour of the seed coat directly with the flower colour, so white seeds came from white flowers and brown seeds originated with purple/violet flowers. Here he suggested that brown seeds associated with the coloured flowers were dominant and by doing so unwittingly introduced the concept of linkage of characters carried close together on the same chromosome.

4. Differences in the form of the ripe pod. Two clear and easily identified forms were described, either inflated, like a pea pod, or contracted about the seeds and deeply wrinkled. It was the inflated pod which was dominant when inherited.

5. Difference in the colour of the unripe pods. Broadly speaking they can be either green or yellow, green being dominant. There was also a variety that had a brownish-red pod that turned blue on ripening which Mendel started to work on the year before these results were presented at Brn, but he either did not pursue his work or it was of little interest as it was never published.

6. Differences in flower position. The two positions described are either distributed along the length of the stem (axial) or being clustered at the top of the stem. It is the axial distribution that is along the stem which turns out to be dominant.

7. Differences in the length of the stem. These are either about 50 cm or about 2 m, the variation which was found being constant for each variety. Mendel noted that while the long stem was dominant, the progeny of the crosses usually exceeded the length of the parental stems, a form of hybrid vigour.

It was the skill of Mendel that he could distinguish the characteristics of his plants consistently. As any plant grower or user of a dichotomous key will be aware, what is written on paper is not always so easy to interpret in the field. There are also very practical problems of which Mendel was aware caused by rogue insects cross-pollinating randomly. He suggested that the main culprit of aberrant pollination was a beetle, *Bruchus pisi*, now *Bruchus pisorum*, and later observers thought swarming thrips may also have caused problems.

With the development of his ideas and from his experimental data, we can see that although chromosomes were never even alluded to, with hindsight they are an essential part of the story. It is only with chromosomes that we can make crosses which are truly independent. Of course, at this stage there was no hint that chromosomes were any part of the story of inheritance, and certainly there was no clue that chromosomes were the massive composite molecules that carried and controlled the genes.

By using such a small number of traits which resulted from well-defined single genes, the basis of inheritance was clearly laid down. But more than that, the manner in which genes, and by implication chromosomes, are transmitted from generation to generation was contained within the work. Mendel's notions of inheritance would fit into the overall picture of genetics as it became elucidated with knowledge of chromosomes. The manner in which Mendel analysed his data also introduced a new method of looking at breeding results by the systematic use of statistics in genetics.

We cannot know whether or not Mendel had any insight into his work as having greater implications than just plant breeding. Whether he did or did not, he certainly made use of his extensive knowledge of plant breeding to develop ideas that would be far more difficult, if not impossible, to do using an animal model. This is not to say that it would not be possible to show single-gene inheritance with an animal system, but it would be a far more specialist job. This was demonstrated by T. H. Morgan with *Drosophila* later on, while William Bateson had suspected this to be the case with his work on poultry. It was the work he had carried out on poultry, showing the discontinuous nature of inheritance in some traits that led Bateson to quickly recognise the significance of Mendel's work (Figs. 3.2 and 3.3).

In the meantime there was a discontinuity between plant breeding as described by Mendel and cellular biochemistry looking at nucleic acids and chromatin. Quite simply there was no apparent association between inheritance and biochemistry. This was an attitude which was going to dog the development of genetics for decades.

However important it was felt that nuclear chromatin was, and it was recognised as being an integral and important part of cell makeup, it remained impossible to be sure where it stood in terms of heredity, or even if it did. Simultaneously, it was impossible to determine what form the material of heredity took. There must be some agent of heredity that was certain, but tracking it down was going to be an exciting and long-winded process. It was known that chromatin was made up of protein and DNA, but could either of these carry the information to make anything as complicated as an animal, let alone a human?

Interestingly, during the time that Mendel was carrying out his detailed analytical work, others were dealing with genetic diseases with no known cause and in some cases not even known to be inherited. This was all part of the burgeoning landscape of diagnostic medicine, which was developing from empirical medicine based on simply treating symptoms, towards trying to understand and treat underlying causes of disease. So it was in the nineteenth century that the stethoscope as a diagnostic tool was developed by Théophile Laënnec in France, and Sir Thomas Clifford Allbutt developed the clinical thermometer. Prior to this taking a person's

temperature involved the use of a long thermometer that took many minutes to register a temperature.

Into this arena came John Langdon Haydon Down. John Down was born in Cornwall, UK, in 1828, and his working life started there when he was apprenticed to his father who was an apothecary. At that time the apothecary was more than a simple mixer of herbs; he was a medical practitioner in his own right. This apprenticeship lasted only until John went to London and started working for a surgeon on the Whitechapel Road. While he was there, he gained some insight into chemistry and physiology which led him to take up medicine at the Royal London Hospital in 1853. He did very well here, and somewhat unusually for a high flier, he

Fig. 3.3 Thomas Hunt
Morgan. Photograph taken
from the John Hopkins
yearbook of 1891

became the medical superintendent at an institution for what would then have been known at that time as the mentally defective. This institution was the rather unappetizingly called the Earlswood Asylum for Idiots. This would have brought him into contact with many patients, some of whom would have been trisomic for chromosome 21. Needless to say, Down would not have known the genetic basis for this condition; what he would have recognised, though, was the uniform nature of the symptoms found in this group of patients (Down 1866).

It was the cataloguing of groups of patients by facial appearance that was referred to in his paper of 1866 where he suggested it was possible to classify various conditions using ethnic characteristics. This is why what we now refer to as trisomy 21 was labelled with the term mongoloid idiocy as a broad umbrella for a group of distinctive patients. This was not to denigrate any race; it merely reflected his perception of similarity between these patients and a known ethnic group. In fact Down was a liberal thinker and used the self-same arguments to rule out any of the ideas of slavery based on one race being superior to another and argued strongly for the unity of the human race.

By observation and differential diagnosis, Down described his eponymous syndrome. The details were presented in 1866 as a London Hospital Clinical Lecture Report with title.

OBSERVATIONS ON AN ETHNIC CLASSIFICATION OF
IDIOTS.

By J. LANGDON H. DOWN, M.D., Lond.

It is interesting that Down's syndrome was sufficiently uniform to be recognisable without recourse to genetic analysis, a situation that was to be repeated in several other conditions involving large-scale chromosome changes. Originally Down's syndrome was referred to as Mongolism or Mongoloid idiocy, which in the middle of the nineteenth century was an acceptable phrase. Nowadays, it is most often referred to as trisomy 21, although this in itself can be misleading. Although it was 90 years after its first description that the human chromosome number was described accurately, it was another 3 years before Lejeune described the numerical aberration responsible. Even then it was not realised that complicated rearrangements and fusions could modify the severity of the condition, and simply counting the chromosomes may give you the right number but the wrong outcome. The description of Down's syndrome having trisomy 21 has had other, minor, implications for geneticists associated with labelling human chromosomes.

In the broad scope of classifying chromosomes, it is normal, regardless of species, to number the chromosomes from largest to smallest, starting with the largest as number one. When the early pioneers of analysis were looking at samples from Down's syndrome patients, they decided it was chromosome 21 which was present as three copies as it was impossible to distinguish between chromosomes 21 and 22 with any certainty. When better techniques were introduced, it was seen that what was now designated chromosome 21 was in fact smaller than chromosome 22. It would by then have caused considerable disruption to alter the designation, and so by convention we keep to chromosome 21 even though it is slightly smaller than chromosome 22. It is not so surprising that Down's syndrome was so clearly defined long before the aetiology was known because the rate of birth at that time was approximately 1 per 650 live births; of course survival was inevitably much lower. Although the natural conception rate will not have changed, live birth rate has with the development of techniques of prenatal diagnosis of this and other conditions.

The situation with Down's syndrome is in marked contrast to another of the only three true autosomal trisomies compatible with life in humans, Edwards' syndrome. This was not recognised as a uniform syndrome until the chromosomal constitution was recognised by Edwards and his colleagues in 1960 (Edwards et al. 1960). Even here, because the chromosomes were countable, but at this time not individually recognisable, the precise designation of chromosome 17 or 18 was not certain. It should be remembered that these solid stained preparations were analysed by researchers still developing techniques. They not only had fewer samples to compare, but they also had no banding techniques to recognise individual chromosomes. A modern cytogeneticist would have seen so many chromosome preparations during their training that they could recognise the individual chromosomes from their size and shape alone. Once described as a definable condition, it became obvious that trisomy 18 came with a number of specific traits. It was most likely that it had not been described in any detail in the past as a syndrome because the birth rate is low, being only about one in 5000 births, and survival rate was also low, only about 10 % surviving more than a year.

A similar situation is seen with Patau's syndrome, also initially described in 1960 (Patau et al. 1960), which involves three copies of chromosome 13, trisomy 13. This, too, has an incidence of about 1 in 5000 births with a very poor survival rate. It was only with cytogenetic analysis that it was determined that these rare births formed a uniform syndrome. Survival of these autosomal trisomies reflects the importance of the chromosomes involved. For some, like chromosome 21, it is both very small and carries a limited number of genes; for others, such as chromosomes 13 and 18, the number of genes may be limited, but just as importantly the genes which are present tend not to take part in the most important aspects of cellular housekeeping. Simply put, if a gene controls hair colour in some way, the importance to the individual is limited. If a gene codes for a histone protein involved in chromatin structure and function, then this is vital for the survival of the individual.

As well as the idea that a single entire additional autosomal chromosome can cause devastating changes in the physiological and biochemical balance of an organism, it was quickly apparent that most chromosomes were too important to be viable as additional extra copies. In contrast to this, very occasionally a triploid individual will be born alive. They do not survive for more than a few hours, but with a total complement of 69 chromosomes, it is surprising they survive to term at all. Most triploids spontaneously abort early on in gestation, but the ones that are live born imply that a balance between various genetic components is as important for survival as absolute genetic content.

The discovery during the 1960s of other more specific conditions, such as Wolf-Hirschhorn syndrome (Wolf et al. 1965; Hirschhorn et al. 1965), the deleted short arm of chromosome 4 in 1965 and Cri du chat (deleted short arm of chromosome 5) in 1963 (Lejeune et al. 1963), unified the ideas of chromosomes being intrinsically important.

It is true to say that more or less any small deletion or addition to any autosomal chromosome can be found, and the catalogue of such deletions and duplications that have been found at some time or another in the last 60 years is immense. It is also true to say that none of the described conditions are positive in their attributes to the individual. The important thing here is the accent on autosomal chromosomes because the sex chromosomes are a special case. They have their own control mechanisms to manage dosage control which ameliorates many of the problems found with changes in autosomes.

The big gap between the diagnostic work of Down and the ability to use knowledge of chromosome numbers as a diagnostic tool is at least in part because there was no link between the three major areas of genetics at the time. Diagnostic cytogenetics was not even thought of and only very little was realised about diseases running in families. Mendelian genetics did not come to the fore until the original papers were translated and more widely disseminated, and last but by no means least, it was only just becoming apparent that chromosomes were not only omnipresent in cells but that they were essential. Tying these together was going to be quite a difficult job, and certainly in the initial phases, it would be botanists and zoologists that made the running.

By the end of the nineteenth century, several strands of investigation were coming together, but the links between them were in some cases not understood and in others not even known. Some idea that chromosomes were important, essential even, had developed and Mendelian genetics was being accepted. There was also a nascent idea of genetic disease, although even with the clear delineation of Down's syndrome, no chromosomal association was made. Joining up the

Fig. 3.4 Edouard van
Beneden photographed about
1910

disparate ideas was going to be a long job and would involve animals far removed
from humans.

The first step towards understanding the vital nature of chromosomes, though not
the chemical nature of inheritance, came with two scientists, one Belgian and one
German. The Belgian was Edouard van Beneden; born in 1845 at Liege, he took
over teaching zoology from his father at Liege in 1870. His work was largely
unpublished before his death, although his lecturers were based on his research, so it
was disseminated amongst his students if not the wider scientific community.
Broadly speaking his most significant work was carried out in 1887 when he
demonstrated that chromosome numbers were constant in all but the sex cells of
an organism (Van Beneden 1870, 1883, 1887). This is reviewed in Hamoir (1992).
His work was carried out using cells from an intestinal parasite of horses which has
four chromosomes in somatic cells and two chromosomes in the sex cells. As a
zoologist Beneden would have appreciated the confusion that the various name
changes of the species he worked on would have. It was originally called *Ascaris
equorum*; this was then called *Parascaris megalocephala* but is now known as
Parascaris equorum (Fig. 3.4).

The German scientist was Theodore Boveri (1862–1915) from Munich where he
originally studied history and philosophy before turning to science. Boveri taught
zoology, first at Munich and then at Würzburg. He not only confirmed the work of
Beneden but extended it by examining other species. With this work he ascertained
to his own satisfaction that the number of chromosomes in a species was constant
and species specific. He was not correct in all of his analyses, which reflects the
technological limitations of his work. He examined many different species, some
being given a chromosome number close to the actual one but not quite correct.
This is the case with his estimate of the number of chromosomes in humans. Boveri
came up with the number 48. Now we know that it is actually 46, but the preparatory techniques available were limited, and we cannot rule out the possibility that
the material he was using was chromosomally abnormal anyway. The idea that
48 was the normal chromosomal complement of man must have taken root at this
time as it was constantly reported over the next 50 years. It was only when new

Fig. 3.5 Theodor Boveri

techniques broke through the limitations of serial sections and squashes that a true modal number was arrived at (Boveri 1902; Fig. 3.5).

Although Boveri died in 1915, his slides were preserved at the university in Würzburg. It was thought that they had been destroyed during the Second World War, but certainly for some of his material, this was not the case. In 2014 it was reported that some of his original material had been rediscovered undamaged at the university. One of the organisms that Boveri investigated was larvae of sea urchins. One of the remarkable aspects of sea urchins, which form the taxonomic class Echinoidea, is their development. Although we often think of extreme metamorphosis being the province of insects such as the Diptera, Lepidoptera or Coleoptera, the metamorphosis that the sea urchin larva undergoes is just as spectacular. Some transitions can take place quite quickly which implies high rates of cell division and consequently a supply of visible chromosomes in the preparations. It is true that sea urchins tend to have large numbers of small chromosomes, but the eggs do have one invaluable attribute; they can be surrounded with a lot of sperm which can result in fertilisation by two sperms. This results, as Boveri noted, in abnormal mitosis with cells receiving either too many or too few chromosomes. That in itself would perhaps not have been significant, but he also noted that the embryos which had abnormal numbers of chromosomes, having been fertilised by multiple sperm, develop abnormally and do not usually survive. So his conclusion that the right number of chromosomes, whatever it was, was important for normal development was well founded and turned out to be absolutely correct.

The work by Boveri was complemented by the work of an American scientist W. S. Sutton who, in 1902, published a remarkable paper which was the first time that the laws of segregation as suggested by Mendel could be directly correlated with chromosomes (Sutton 1902). Walter Sutton had a short career; he was born in 1877 and died in 1916 aged 39, due to problems associated with an appendicitis.

Sutton worked on grasshopper testes which was an excellent choice of organism because they have large and readily visualised chromosomes. There is more than that, though, to this experimental organism. The sex chromosome is distinguishable from the autosomes. It had already been suggested by C. E. McClung (1901, 1902, 1909) that the apparently supernumerary chromosome found in grasshoppers may be in some way responsible for determining sex. What Sutton did was look at spermatogonia which contained the diploid number ($2n = 23$) and then try to work out whether the size of the chromosomes was constant. This would have been an extremely difficult investigation because the variation between cells in observed size would be considerable. It was in effect a very difficult practical problem to work out a method of doing this. As part of his investigation, he noticed that within any given cell it was possible to pair up the chromosomes into eleven distinct couples with the accessory chromosome left over. This had significance in basic understanding of chromosomes and their importance.

By studying the various stages of meiosis in grasshopper testis from spermatogonia to sperm, it looked very much as though the chromosomes that became visible and paired up for the first cell division paired up as matches for each other. They were pairing with their homologue as recognised by size and shape. Further than that, when the number of chromosomes was reduced in meiosis, it did not happen by chance; it was always one of each homologous pair which went to each daughter cell.

Sutton suggested, correctly, that exactly which chromosome of each pair, paternal or maternal in origin, went to the daughter cells was random. Similarly it was random as to which cell the accessory chromosome went to, so some would have it and some would not. The sperm cells that did have it were destined to fertilise an egg that would produce a female and those without would produce a male offspring. Of course this is different to the human sex chromosome system which involves two sex chromosomes, but the implication for sex chromosomes and sex determination was obvious (Fig. 3.6).

Fig. 3.6 Walter Sutton

It was a paper that Sutton published a year later in 1903 (Sutton 1903) that made the leap from observation to interpretation that explained inheritance by tying together the behaviour of chromosomes and the breeding experiments of Mendel. Sutton envisaged units of heredity being carried on chromosomes. By thinking of Mendel's characters as being on separate chromosomes, he could comfortably demonstrate the same results not from breeding experiments but by direct observation of chromosomes. It would seem that Sutton had originally thought that maternal and paternal chromosomes operated within the cell independently, segregating and operating as though they were two completely independent sets. It was with the rediscovery of the work of Mendel that it occurred to him that there was a distinct correlation between his work and the inheritance of traits if the assortment of maternal and paternal chromosomes was random; this introduced the element of chance into cell division which can be interpreted using probability theory.

During this period when the association between what we now think of as classical genetics, that is, Mendelian inheritance, and chromosomes was recognised, it would be easy to assume that the relationship would be causal. It should be remembered though that making this assumption would have left too many unanswered objections to a direct relationship between chromosomes and inheritance. It would be easy to see a relationship and make one assumption too many. It is true that the inheritance of discrete characters as described by Mendel followed the movement of chromosomes as described by Sutton, but this did not automatically mean that there was a demonstrable cause and effect. One aspect of chromosomes that did not escape the attention of those willing to presume they carried units of heredity was that there were more things which needed coding for than there were chromosomes, so each chromosome would have to carry a great deal of information.

For some scientists the relationship between Mendelian genetics and chromosomes was incompatible. Probably the most well known of the chromosome dissenters were William Bateson and Thomas Hunt Morgan. It may seem odd to include Morgan here, but he certainly had some reservations regarding the relationship between Mendelian inheritance and chromosomes. Like Bateson, Morgan accepted Mendelian inheritance but could not understand why the apparent low number of chromosomes relative to the number of inherited traits did not result in lots of characters being inherited en bloc.

While Bateson was a staunch advocate and supporter of the work of Mendel (Bateson 1901) and introduced the word 'genetics' into the vocabulary, it was Morgan who started a systematic line of research into inherited traits. His species for study was primarily *Drosophila melanogaster*, often just referred to as fruit flies or by the genus *Drosophila*. Bateson did eventually give some credibility to the idea that chromosomes were the transporter of genes, although it is said that he never fully accepted all the ramifications of the idea (Bateson and Saunders 1908; Bateson and Punnett 1908). On the other hand, Morgan not only tested the idea but demonstrated it unequivocally.

Drosophila is a genus of small dipteran flies, and like all dipterans they have only a single pair of wings, the second pair being highly modified. *Drosophila* has a relatively fast breeding cycle which means that although holometabolic they can pass through a generation in under 2 weeks. Besides this even though they are small, being only a few millimetres long, they do have a large number of easily recognisable characters, such as eye colour and bristle numbers on specific plates. This makes demonstrating Mendelian inheritance in these small organisms possible. As experimental organism, they also have a major advantage in their chromosomes and the easily recognised sex chromosomes. It is *Drosophila* which is instantly thought of when geneticists refer to polytene chromosomes, although these appear in many other fly species and many other organisms including amphibians. It was not initially realised how useful the giant polytene chromosomes of the salivary glands were going to be in research. Morgan started work on *Drosophila* in 1907 after he had accepted the chromosome theory of inheritance. Morgan started out by investigating linkage, the phenomenon of traits being transmitted together rather than assorting independently. It was originally Sutton's hypothesis that these traits were carried on the same chromosome, so to test this he developed his breeding programme. Morgan also reasoned that genes were physical entities that could be precisely located on a chromosome. There was little or no evidence for this as it could equally have been a number of individual things working together to create a specific trait, but like all good scientists, investigations start with a testable hypothesis.

There was one particular problem with the idea of linkage; occasionally linked genes would suddenly appear to separate, effectively becoming unlinked. The combined idea that genes were single entities and yet they could, by implication, move between two chromosomes, even if they were homologues, was radical and for most of the time difficult to swallow without some pretty strong evidence. Before Morgan provided his own results on crossing over and unbeknownst to him, F. A. Janssens had been looking at meiosis and noted the apparent break and rejoin process which is now what we know as crossing over (Janssens 1909). He suggested that the breakage of two chromosomes simultaneously in the same place may well result in rejoining of different homologues. Although cytological proof was not going to be available until 1931, by embracing the idea of crossing over, swapping versions of the same gene, between homologues, Morgan managed to retain the idea of genes having a fixed position on a chromosome (Fig. 3.7).

While Morgan was working on fruit flies, he discovered a white-eyed mutant, contrasting to the red-eyed wild type. This seemed to trip up the normal mode of inheritance, until it was realised that the eye colour was sex linked (Morgan 1919, 1926). By finding more of these sex-linked mutations, Morgan and his student Sturtevant made a remarkable discovery (Sturtevant 1913). They had already shown that genes would become unlinked at a constant rate for any pair of genes and that it varied from gene pair to gene pair. By marking the rate of unlinking that took place between different genes, between pairs of different genes, it was possible to say what the relative position was of any given gene compared to another. Although it is relative, the recombination event frequency can be converted directly

Fig. 3.7 Frans A. Janssens

into a relative distance. So if two genes recombine at a rate of 10 % or 0.1, it is expressed as 10 centimorgans. Interestingly crossing over is not uniform along the length of a chromosome so a linkage map will be distorted towards the centromere and the telomeres of the chromosome when compared to a physical map.

Using this technique, it was possible for Morgan's team to produce the first chromosome map in 1911. This was composed of five sex-linked genes in *Drosophila*. This number had grown to over 2000 by 1922 and now represented all four of the fly chromosomes. In 1933 Morgan received a Nobel Prize for his work.

There was at this time a gap between the functional descriptions of inheritance, as produced by Mendel, Sutton, Morgan and other scientists working in the early part of the twentieth century and the physical and chemical basis of it, which was being investigated by biochemists.

The exact chemical nature of inheritance was going to be difficult to elucidate, but the chemical structure of chromatin was going to be easier to work out. The gap between chemical structure and functional structure was simply one of analysis. The chemical structure of any organic molecule can be solved by a reductionist approach. By that we mean that chemical analysis will yield an empirical value of the constituent parts. We can even give a rudimentary idea of the empirical formula for a human.

$$H_{15750} \, N_{320} \, O_{6500} \, C_{2250} \, Ca_{63} \, P_{48} \, K_{15} \, S_{15} \, Cls_6 \, Mg_3 \, Fe_1$$

This may seem a ridiculous idea and it is certainly an extreme reductionist viewpoint, which may not even be very accurate, but it does show that while the chemical analysis may be possible, it tells us nothing about the molecular structure or function of the item described. The first thing, then, was to determine where the material of heredity resided. Walther Flemming (1843–1905) was a zoologist specialising in cytology; in fact he was a pioneer in this field. It was Flemming who described the nuclear changes during cell division using newly available aniline dyes (Flemming 1882). He demonstrated that cells contained areas that

Fig. 3.8 Walther Flemming, c.1905

absorbed basophilic dyes, which he described as chromatin. By careful observation, he recognised the association between chromatin, the nucleus and thread-like structures which Waldeyer-Hartz would name as chromosomes. It is interesting to note that although originally in German, the seminal paper of Waldeyer-Hartz (1888) was translated and reprinted in English a year later (Waldeyer-Hartz 1889). Even for the time, this was a long paper at about 120 pages. There was, during the nineteenth century, less pressure on brevity in published works of this type. This is demonstrated by the way in which the word 'chromosome' was introduced on page 181 of the journal. Waldeyer-Hartz wrote 'I must beg leave to propose a separate technical name 'chromosome' for those things which have been called by Boveri "chromatic elements"'. This translation appeared in the *Quarterly Journal of Microscopical Science* which changed its name in 1966 to the *Journal of Cell Science* (Fig. 3.8).

It was during this period, before the widespread recognition of the work by Mendel and before Sutton had made his association between inherited characters and chromosomes, that Albrecht Kossel started investigating the recently isolated substances called nuclein (Kossel 1884a, b, 1885). Kossel (1853–1927) worked on many aspects of cell biochemistry and was awarded a Nobel Prize in 1910 for his work on cellular chemistry. In 1879 he started investigating nuclein which he first demonstrated to be a composite of protein and nucleic acid. Between 1885 and 1901, he and his students discovered the key components of adenine, cytosine, guanine, thymine and uracil, which replace thymine in RNA. It is the last of these which gives a clue that he was not just analysing DNA but was also analysing RNA. This fact would not be conclusively known until the American Phoebus Levene (1869–1940) furthered the precision of the analysis. Kossel also ascertained there was a phosphoric acid and a sugar. He did recognise that the bases AGCT and U were from either of two groups, purines which he named adenine and guanine and pyrimidines which he named cytosine, thymine and uracil.

Levene made great strides in the understanding of nucleic acid (Levene 1932). He managed to show that the sugars Kossel had not identified were of two types, deoxyribose and ribose, and that the ribose version contained AGCU, while the deoxyribose contained AGCT. By further investigation, he showed that the nucleic acid molecule was made up of discrete units. These were a phosphoric acid molecule, a sugar, either deoxyribose or ribose and a base, which could be either a purine or a pyrimidine. If the sugar was ribose, then one of the pyrimidines would be uracil, but not thymine, and conversely if the sugar was deoxyribose, one of the pyrimidines would be thymine and not uracil. The other pyrimidine, cytosine, and the purines remained unaltered regardless of which sugar type was involved.

With the work of Sutton and Mendel showing that inherited characters followed distinct and precise lineages and were associated with the physical structures observed as chromosomes, the question of what was the nature of the units of inheritance arose. Chemical analyses of nucleic acids by Kossel and Levene amongst others demonstrated that they were chemically very simple in comparison with proteins, which were always present in the nucleus. As a consequence of this observation, it was assumed that something as simple as the nucleic acids could not possibly code for something as complicated as a person, so surely it must be the associated protein. There were some complications with this argument as it had been noted by Kossel that in sperm heads the amount of nucleic acid present was considerable, but there was a paucity of associated protein. Nonetheless, it was very difficult to see how a simple molecule, nucleic acid, could possible carry any information of any significance.

There was, at this time, a lack of adequate information and understanding about the nature of genetic inheritance. The problems could be easily stated, and the broad functional descriptions of what was known could also easily be stated. The problem arose because joining up the two areas of work was impossible without information to back up or refute hypotheses. This was a time when biologists were facing a situation which they had never faced before. Until the nineteenth century and the advent of microscopes which revealed things which had to be explained, like cells and microbes, biologists were observers of the macro-world. Now they had to deal with the invisible, and they had to accumulate data from unseen things which they did not know anything about to explain the world they could see. Joining up the two arenas was going to be difficult. This is one of the reasons that the input of chemists was going to be essential to reach a successful conclusion regarding the structure and function of the gene. At the same time the investigation of chromosomes, which could be seen, would continue in the hands of geneticists and biologists. It is interesting that this dichotomy would exist for a long time, with each group often paying little attention to the other.

References

Bateson W (1901) Experiments in plant hybridization. J R Hortic Soc 26:1–32
Bateson W, Punnett RC (1908) The heredity of sex. Science 27:785
Bateson W, Saunders PR (1908) Confirmations and extensions of Mendel's principles in other animals and plants. Report to the Evolution Committee of the Royal Society, London

Boveri T (1902) Über mehrpolige mitosen als mittel zur analyse des zellkerns. This was translated and republished as- Boveri T (1902) On multipolar mitosis as a means of analysis of the cell nucleus. Found Exp Embryol 1964:74–97

Down JLH (1866) Observations on an ethnic classification of idiots. Clin Lect Rep Lond Hosp 3:$ 32#259–262

Edwards JH, Harnden DG, Cameron AH, Crosse VM, Wolf OH (1960) A new trisomic syndrome. Lancet 275(7128):787–790

Flemming W (1882) Zellsubstanz, kern und zelltheilung. Vogel, Leipzig

Hamoir G (1992) The discovery of meiosis by E. Van Beneden, a breakthrough in the morphological phase of heredity. Int J Dev Biol 36(1):9–15

Hirschhorn K, Cooper HL, Lester Firschein I (1965) Deletion of short arms of chromosome 4–5 in a child with defects of midline fusion. Hum Genet 1(5):479–482

Janssens FA (1909) La Théorie de la Chiasmatypie. Cellule 25:389–414

Kossel A (1884a) Üeber Guanin. Z Physiol Chem 8(5):404–410

Kossel ALKML (1884b) Ueber einen peptonartigen Bestandtheil des Zellkerns. Z Physiol Chem 8(6):511–515

Kossel A (1885) Ueber eine neue Base aus dem Thierkörper. Ber Dtsch Chem Ges 18(1):79–81

Lejeune J, Lafourcade J, Berger R, Vialatte J, Boeswillwald M, Seringe P, Turpin R (1963) 3 cases of partial deletion of the short arm of a 5 chromosome. C R Hebd Seances Acad Sci 257:3098

Levene PA (1932) Nucleic acids. Am J Med Sci 183(6):855

McClung CE (1901) Notes on the accessory chromosome. Anat Anz 20:220–226

McClung CE (1902) The accessory chromosome—sex determinant? Biol Bull 3(1–2):43–84

McClung CE (1909) Cytology and taxonomy, vol 4, no. 7. University of Kansas, Lawrence, KS

Mendel G (1866) Versuche über Pflanzen-Hybriden. Verh Naturforsch Ver Brünn 4:3–47

Morgan TH (1919) The physical basis of heredity. JB Lippincott, Philadelphia

Morgan TH (1926) The theory of the gene. Yale University Press, New Haven, CT

Patau K, Smith D, Therman E, Inhorn S, Wagner H (1960) Multiple congenital anomaly caused by an extra autosome. Lancet 275(7128):790–793

Sturtevant AH (1913) The linear arrangement of six sex-linked factors in *Drosophila*, as shown by their mode of association. J Exp Zool 14:43–59

Sutton WS (1902) On the morphology of the chromosome group in *Brachystola magna*. Biol Bull 4(1):24–39

Sutton WS (1903) The chromosomes in heredity. Biol Bull 4:231–251

Van Beneden E (1870) Recherches sur la composition et la signification de l'oeuf. Mémoires couronnées et Mém. des savants étrang., publiées par l'Académie Royale de Belgique

Van Beneden E (1883) Recherches sur la maturation de l'oeuf, la fécondation et la division cellulaire. Arch Biol 4:610–620

Van Beneden É (1887) Nouvelles recherches sur la fécondation et la division mitosique chez l'Ascaride mégalocéphale. W. Engelmann, Leipzig

Waldeyer-Hartz W (1888) Über Karyokinese und ihre Beziehungen zu den Befruchtungsvorgängen. Archiv für mikroskopische Anatomie und Entwicklungsmechanik 32:1–122, paper in German, translated below

Waldeyer-Hartz W (1889) Karyokinesis and its relation to the process of fertilisation. Q J Microsc Sci 30:159–281, this is a translation of the above paper

Wolf U, Reinwein H, Porsch R, Schröter R, Baitsch H (1965) Deficiency on the short arms of a chromosome no. 4. Humangenetik 1(5):397

Chromosomes as the Carriers of Heredity

4

Although the work of geneticists, breeders and cell biologists had transformed our understanding of chromosomes and the nucleus as carriers of heredity, questions still remained. By 1920 it was clear that some things were known for certain, some things were thought to be true and there was a lot that just was not known. There was by this time an acknowledgement that chromosomes were crucial in inheritance of genetic traits, in some unknown way carriers of genes. It was also known that they were made up of deoxyribonucleic acid (DNA) and various proteins. What was not known was how many chromosomes there were supposed to be in a human cell and how the genetic information was carried or contained. It was going to take a good few years to determine that the material of heredity was DNA and even longer to be sure what the chromosome number was; in the intervening years, there would be no shortage of ideas.

Even as far back as 1926, it had been vaguely suggested that assumptions regarding the central importance of proteins should not be assumed through lack of contrary evidence. This was a straightforward suggestion that without experimental evidence nothing could be assumed to be true simply by virtue of prejudice. On the other hand, there was a sound logic to support the idea that protein was the material of importance in chromosomes, the very basis of inheritance and even possibly the material from which genes were composed. These ideas came about with amino acids first being isolated and described chemically in 1806 when two French chemists Louis Nicolas Vauquelin and Pierre Jean Robiquet isolated an extract from asparagus (Vauquelin and Robiquet 1806). This was the first of the amino acids to be determined and it is no surprise that it became known as asparagine. Vauquelin was a chemist of considerable skill, being the discoverer of chromium in 1797 when he examined the rare mineral crocoite. This is lead chromate, $PbCrO_4$, from which he extracted chromium trioxide, and then after extensive heating, he drove off the oxygen to be left with the metal. Only a year later, he studied the minerals beryl and emerald, having been suggested to him that since the crystal forms were identical, they were probably chemically identical as well. Vauquelin not only proved this to be correct but that the green colour of

© Springer International Publishing Switzerland 2016
W.J. Wall, *The Search for Human Chromosomes*,
DOI 10.1007/978-3-319-26336-6_4

Fig. 4.1 Louis Nicolas
Vauquelin

Fig. 4.2 Pierre Jean
Robiquet c.1830

emerald is due to traces of chromium. Although he realised that these two minerals, which we now know are beryllium aluminosilicate, contained another new metal, beryllium, he did not succeed in isolating it (Figs. 4.1 and 4.2).

More amino acids were isolated over the next decades, the second one which was described was cystine in 1810. This was discovered by William Wollaston, although it was not appreciated as a constituent part of proteins until 1899. It is interesting to note that he refers to it as cystic oxide extracted from a urinary calculus, which in this situation means a bladder stone (Wollaston 1810). Before turning his analytical skill towards amino acids, Wollaston had been involved in methods of purifying platinum, during which he had discovered two new elements: palladium in 1802 and rhodium in 1804. The next time that amino acids are described is in 1820 when the French chemist Henri Braconnot described glycine and leucine (Braconnot 1820; Fig. 4.3).

It was, therefore, long before nucleic acid was described that the chemistry of proteins was slowly being unravelled. That is not to say that it was immediately

Fig. 4.3 Henri Braconnot

understood what the significance of amino acids were. It was soon realised that enzymatic digestion of proteins would result in amino acids, but it was not until 1902 when Emil Fischer and Hofmeister independently grasped that proteins were constructed by stringing together chains of amino acids that the essential nature of proteins was understood (Hofmeister 1902; Fischer 1906; Fruton 1985). Since there were about 20 amino acids which could be extracted from proteins, there was an incalculable variety of proteins available, without even beginning to tap into conformational folding. Interestingly, Emil Fischer received the second Nobel Prize to be awarded in chemistry in 1902, being cited for his work on synthesis of sugars and purines (Fischer 1902; Fig. 4.4).

So more or less at the turn of the nineteenth century, proteins already had a reputation for complexity and variety and were known to be a major part of the structure of chromatin. Against this background, DNA looked to be relegated to a subsidiary role as some sort of structural hanger-on. Into this argument some formidable chemists became convinced of the impossibility of DNA being a code carrier. Surely a complicated three-dimensional structure—like a human—requires a lot of coding and control to make it work properly, so it must take a complicated structure like a string of amino acids wound up into protein to do the job. The concept of coded information was a relatively new one during the nineteenth century and not widely appreciated in its implications. Morse code had been devised in the 1840s, although it was not until the 1890s that it became an essential part of radio telegraphy before voice transmissions became commonplace. Until that time it had never been necessary to have systems in place where a message could be sent without ambiguity, so the idea of a simple molecule coding for a complex structure was not easily understood. Besides, if you have a coded message, it has to be translated in some way before it can be acted upon. The idea of DNA taking on this role was hard to comprehend, but proteins in all their majestic variety were already there for all to see—surely they could simply replicate themselves and perform all of the functions of heredity.

This was poignantly put in a paper published in the Transactions of the Faraday Society in 1935 where it was said that:

> If one assumes that the genes consist of known substances, there are only the proteins to be considered, because they are the only known substances which are specific to the individual.

Fig. 4.4 Evolution of a
journal. Founded in 1879
Annales de Chimie lasted
until 1815 when it became
*Annales de Chimie et de
Physique*. In this form it
carried on for almost a
century until 1914 when it
divided to become two
separate journals, *Annales de
Chimie* and *Annales de
Physique*

ANNALES DE CHIMIE.

31 *Mars* 1815.

MÉMOIRE

Sur la nature des corps gras ;

Par M. Henri Braconnot,

Professeur d'Histoire naturelle , directeur du Jardin
des plantes , et membre de l'académie de Nancy.

Lu à la société des sciences, arts, agriculture et belles-
lettres de Nancy, le 9 février 1815.

Jusqu'à présent les chimistes ont consi-
déré les graisses des êtres organisés comme
étant formées d'une seule et même substance,
ayant les mêmes propriétés essentielles , et
ne différant d'une façon marquée que par sa
consistance plus ou moins ferme : delà , ces
dénominations de suif, axonge, moelle ,
graisse, etc. , admises par les anciens. Cette
consistance de la graisse varie en effet d'une

Tome XCIII.   15

This takes the idea of complex molecules being necessary to produce complex organisms one step further by suggesting that it requires a molecule or molecular system of enormous variety to be capable of creating the variation seen between individuals, not just between species. One other idea arose from P. A. Levene which was based on the observation that there appeared to be only four different bases in DNA (Levene 1919). Levene was working in the USA and had considerable experience in handling and analysing DNA. He made an influential suggestion which was basically that since there was such a limited scope for variation in the arrangement of four bases, it stands to reason that they would be constructed in repeated tetrads. This tetranucleotide hypothesis had very little evidence in support of it but nonetheless gained widespread acceptance. This was understandable since the alternative idea that four bases coded for life was simply too fanciful. It would require too many additional parts to make the system work. It is much easier to simply say that complicated bodies are coded by complicated proteins. This, of course, harkens back towards the homunculus idea of a complete human in miniature being present in a sperm head, by implicitly assuming that all necessary proteins for human structure and function were present at conception. Difficulties such as lack of protein compared to DNA in sperm tended to be skated over. Because there really was no consensus amongst contemporary geneticists of the day as to whether it was protein or DNA, it is usual to find no explicit reference to the mechanics of inheritance in textbooks of the day. Beyond noting that the rules of Mendelian inheritance can be demonstrated microscopically by watching chromosomes, questions of gene structure are avoided. This lack of engagement with the debate over gene structure was also partly because genetics was still being taught primarily as Mendel would have known it while analysis of the contents of the nucleus was being carried out down the corridor in a different department, usually chemistry. In 1934 Crane and Lawrence wrote in *The Genetics of Garden Plants* (Macmillan and Co.):

> The precise nature of the gene is at present unknown but it seems probable that it is a highly complex molecule or group of molecules with the characteristic power of self reproduction.

Steering a narrow line between committing to one side of the argument or the other, they seem to imply a rational agreement with protein being a contender. It has to be emphasised, though, that generally at this time comments in texts were worded in a defensive way. Should the author make a statement in favour of one unproven idea or another which turns out to be incorrect, their reputation may be damaged.

The shift in the collective idea away from protein being the front runner as director of heredity towards a realisation that DNA was the genetic material was a slow process. There were many ambiguous results to be explained, and it was difficult to dismiss protein as the agent of inheritance. The first experiments which laid a foundation for the understanding of DNA as central came in 1928, although the interpretation of these results were tentative, resulting in no solid conclusions. Griffith was looking for a suitable treatment for pneumonia and certainly not any understanding of genetics or inheritance; as a medical officer, he was after a practical solution to the growing problem of pneumonia in crowded cities with appalling air quality. His work was carried out in London for the Ministry of Health where the problem of pneumonia was well recognised as a major killer in the years after the First World War. This was a time of increasing concern in public health as the post WW1 epidemic of Spanish influenza had caused so much misery and fatalities that action had to be taken whenever a perceived threat was found. This was true even though for the most part it was a concern for the number of lost working days rather than for the general health of the population.

Griffiths worked on *Streptococcus pneumoniae* which he had recognised to come in two forms (Griffiths 1928). The first is a bacterium with a polysaccharide coat which causes pneumonia and caused the fatalities, while the second lacks the coat and is killed by the host organism, causing few, if any, symptoms. So one form, encapsulated, is virulent and the other form is not. The structure of the experiment was quite straightforward. Mice that were injected with pneumococci from encapsulated live strains died from the infection. Those injected with live nonencapsulated strains did not. If the encapsulated bacteria were heat killed, the mice also survived. If, however, the killed encapsulated strain was mixed and injected with nonencapsulated but live bacteria, some of the mice died, and encapsulated live bacteria could be isolated from the dead mice. Something was transforming the non-lethal strain into the lethal form. What exactly this transforming principal was remained unknown for some years, although in 1931 Dawson and Sia demonstrated that the transformation from one form of bacterial phenotype to another was possible in vitro. Griffith himself never worked out what the transforming agent was and never saw the significance of his work recognised as he was killed in an air raid on London in 1941. What his work did encourage was an interest in transformation of bacterial strains and directly influenced Oswald Avery to look at the problem in more specific detail.

Transformation became of interest to many groups because it seemed that there was an observation on inheritance that would yield useful knowledge about genetics. By formulating a precise hypothesis that could be tested using an experimental approach, it was recognised that significant strides could be made in answering a fundamental question in biology.

If the transforming element could be chemically separated and demonstrated to alter the gene expression of the bacteria, then that should be the material of heredity. Although it is simply stated, the experiments and interpretation of results had many practical and theoretical hurdles to cover. This task was taken on by a group headed by O. T. Avery working at the Rockefeller Institute. As many research groups have found, specific requirements for equipment were best served by constructing them yourself. So when it became necessary to produce large amounts of bacterial culture, a process not normally done at that time, the ingenuity of the team was put to a very practical purpose. Until the system was geared up, production and extraction of the *Streptococcus pneumoniae* had been carried out in batches which were only between 3 and 5 l. When they wanted to make extractions from their 36 l cultures, a new way of extracting the cells was required as the old way proved unable to replicate the transforming ability on the larger scale. The need for this scaling-up was all due to there only being less than 0.5 g of bacteria in a litre of culture. With the standard laboratory centrifuges, they could separate a litre of culture in an hour into cells and supernatant. What the ingenious scientists did was convert a device designed to separate cream from milk by a process of constant flow. With a few modifications, most notably to stop bacterial aerosols, the device worked well enough for large-scale cultures to become routine. By producing enough material, Avery, McLeod and McCarty finally published their results on transformation in 1944, confirming that it was actually DNA which caused the change and could be regarded as the genetic material (Avery et al. 1944). Once they had produced enough bacteria, they could be lysed by heat shock, heating and freezing, so that the cells were disrupted. The resultant mixture of broken cells and liquid could then be divided into the solids, which constituted the cell walls and bulk of the proteins and the supernatant, which contained the water soluble parts of the cytoplasm. It was this supernatant which contained the transforming material. This confirmation that DNA could transform the serotype of bacteria, changing the genetic makeup of the cells, was not instantly recognised due to the date of the publication, it being during the Second World War. Consequently it remained not exactly ignored but was slow to catch on (McCarty 1985).

There is another reason which might have slowed the process of dissemination and demonstrates the importance of choosing the right vehicle for publication of scientific results. The paper was published in the *Journal of Experimental Medicine*, a journal originating at Rockefeller University. In the same edition, all other papers, except one, were microbiological. The exception was on vitamins and rats. Without the modern indexing systems, a journal at that time relied on subscriptions to disseminate its content, so the process would have been slow anyway, but publishing it in a journal not normally seen by the genetics community would have hidden it from direct view until it was described peer to peer.

It was some time before the idea of DNA being the pivotal element in heredity gained complete acceptance. There had been much embedded philosophical belief in proteins being the only suitable carrier of genetic information. Even 2 years later, in 1946, there were papers voicing doubts as to the efficacy of the 1944 conclusions. Also, there was the problem of cross-fertilisation of ideas. It was then, and to a lesser extent 50 years later, commonplace for groups studying different subjects to ignore other people's results. So a paper in the *Journal of Experimental Medicine* from a group working in bacteriology was not going to be quickly disseminated amongst geneticists, who for the most part preferred to work on multicellular animals and plants where individual characters could be seen in individual organisms.

A few years later in 1952, another well-thought thorough experiment made the belief that protein was the genetic material untenable. It also underlined DNA as the transforming principle, the carrier of heredity. It did not make it any easier to understand how this simple molecule could do so much, bringing forward ideas of codes and encrypted information. The experiment which finally made the difference was carried out by Alfred Hershey and Martha Chase (1952). They were working with a bacteriophage T2 using radioactively labelled elements. An interesting point about this particular line of attack was that it would have been impossible a few years previously as the two isotopes which were going to be used were artificially produced. ^{32}P (half-life 14.3 days) and ^{35}S (half-life 87.1 days) are not normally found in nature but were to become mainstays of biochemical research for many decades after 1950 (Fig. 4.5).

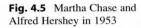
Fig. 4.5 Martha Chase and
Alfred Hershey in 1953

Hershey and Chase grew bacteriophage T2 on bacterial colonies containing ^{32}P
or ^{35}S, knowing that while phosphorus is an essential component in DNA, it is not
found in protein, and sulphur, essential to many amino acids, is not found in DNA.
In this way they produced large amounts of phage with either their protein coat or
their DNA radioactively labelled. Using the labelled phages to infect new colonies
of bacteria that had no radioactive labels themselves, they hoped to demonstrate
that either DNA or protein was essential for inheritance. The colonies infected with
^{35}S-labelled phage produced new phage without labels, the protein having been left
behind outside the bacteria when it injected DNA into the cell. On the other hand,
^{32}P-labelled phage injected the radioactive DNA into the bacteria, and as DNA was
conserved during replication, the radioactive label and the phage which were
produced did contained ^{32}P too. This was a confirmation that DNA was the very
essence of inheritance, not protein, as this was discarded during viral replication.

One of the reasons that DNA as a carrier of detailed genetic information was
thought unlikely was based upon the pervading idea of the tetranucleotide as the
normal construction of DNA. This had been put forward and championed by P. A.
Levene at Rockefeller University, but there was a little more detail in this which
added to the scepticism of DNA as a primary information carrier. This was
associated with the tetranucleotide hypothesis but slightly nuanced. If the base
composition was looked at chemically, then the purines equalled the pyrimidines in
number. Even more accurately, the individual base adenine equals thymine and
guanine equals cytosine.

As a complication, AT does not necessarily equal GC. This relationship between
base equivalents was shown to be universal by Erwin Chargaff in the early 1950s
and laid the foundation of the possibility of a code being present in the sequence of
bases (Chargaff 1950). Chargaff was a Czech who studied extensively in Europe
before moving to the USA where he produced his most well-known work on DNA
while at Columbia University. His analysis was further supplemented by finding

that DNA came in much longer chains than had originally been imagined. This was because the chemical analysis had not taken into account the structures in which DNA was found, the chromosomes, but simply took the analysis as a chemical process. It was the very long pieces of DNA which was going to be crucial to understanding the way in which it worked, not the simple basic chemical structure.

It was, indeed, just this chemical analysis, regardless of context or ancillary controls, which gave us a clear understanding of how it was that a linear strip of bases could be converted into a protein. This was important as it had already been suggested that one gene produced one enzyme, later changed in wording if not principle to one gene one polypeptide. Although it took some considerable time to work out the commaless triplet code which took DNA and processed it into protein, it did demonstrate a very fundamental attribute of universality. This is significant as the more was known about the small scale chemistry of DNA and genetics, the more it became apparent that this is not a species-specific subject. Broadly, what is learnt in one organism is applicable to all others. This is especially useful in the case of humans where as experimental organisms we fail for many reasons, so it is much better for us to learn about genetics in one species and then apply it to ourselves.

When it was accepted as certain that DNA was the carrier of genetic information, attention inevitably moved towards questions of detail regarding structure and function. How could such a simple molecule contain such a wealth of information and how was it constructed so that it was reliably self-replicating? With the work of the various researchers looking at the structure of DNA, resulting in the famous double-helix model, and with the deciphering of the triplet code, an interesting dichotomy became apparent. Between the geneticists looking at Mendelian inheritance and functional genetics in multicellular organisms and the chemists looking at the minutiae of DNA structure and function, there was a massive crevasse.

While chemists analysed the structure of DNA in a very matter-of-fact way with a reductionist view to the structure and function of genetic material and genetics as a whole, geneticists were working rather differently. Genetics was developing as a subject at the hands of biologists and so tended to involve eukaryotes and, more often than not, multicellular ones. This was because determining a functional description of what was happening generally involved looking at inheritance of single characters. To make the most of these Mendelian traits which result in a changed phenotype requires the organism to be large enough to have easily recognisable characters and progeny in numbers that could be counted for statistical analysis. In contrast to this, the analysis of chemical makeup needed large amounts of readily available material. This requirement was most readily available from cultures of simple organisms which were mainly prokaryote bacteria.

With this twin track in genetics going on, there was really a disappointing amount of cross-fertilisation of ideas between what might be called classical genetics and biochemical genetics. There is no doubt that this caused some difficulty in the development of clear ideas of how genes work. Later on it even made for some huge overestimates of the number of genes that are present in the human genome. Almost as soon as the approximate number of bases in the human genome was calculated, the bold statement was made that this represented more than

100,000 genes (Liang et al. 2000). This was quickly accepted and yet at the same time human geneticists were saying that the figure would be more like half that number. This more accurate and smaller estimate was made by working out the average size of genes and then taking into account control regions and non-coding regions as well as repeat sequences and all the other paraphernalia that makes up a genome. In this case the reductionist view of the biochemists was shown to be far off the mark, and although the guess of the genetics community of 50,000 was nearer the truth, that was still far too large. More accurate estimates have brought the number down by a considerable amount (Ewing and Green 2000).

The gulf was exacerbated by the types of journals in which the data was published. Before the advent of journals capable of taking articles on any aspect of genetics, most of the background work appeared in quite specialist journals, usually ones which generally featured articles about the experimental organism rather than the type of investigation being reported. Then, as now, to appeal to the largest audience which might be interested means publishing in the journals with the widest readership. This is why, for example, the Watson and Crick structure of DNA was published in *Nature*. For many of the other seminal research results, they appeared in their own relatively small circulation journals.

Even though the importance of chromosomes in genetics and heredity had been known for a long time, the significance of the knowledge seemed for many to be separated from the importance of the chemistry. In many ways this is a reflection of answering small questions which have a solution rather than asking the big questions which are difficult to formulate in any sensible way, for example, why have chromosomes? Of course it was realised they must serve some function, but exactly what was unknown. Chromosomes have a distinct problem when it comes to being investigated. They are too small to be easily handled and yet they are far too big to be treated as molecules. In many ways they can be visualised as gigantic macromolecules made up of a DNA polymer and associated protein.

With the understanding that chromosomes were not merely important to Mendelian genetics, but actually the reason that heredity works the way it does, with segregation and linkage all being a part of the structure and function of chromosomes, came another sort of enquiry. As chromosomes were looked at in more detail, generally to start with in plants, it became apparent that they were not simple strips of chromatin; there was far more to them than that, but they were being elusive and not yielding to simple analysis.

References

Avery OT, MacLeod CM, McCarty M (1944) Studies on the chemical nature of the substance inducing transformation of pneumococcal types induction of transformation by a desoxyribonucleic acid fraction isolated from pneumococcus type III. J Exp Med 79(2):137–158

Braconnot HM (1820) Sur la conversion des matières animales en nouvelles substances par le moyen de l'acide sulfurique. Ann Chim Phys Ser 2:13

Chargaff E (1950) Chemical specificity of nucleic acids and mechanism of their enzymatic degradation. Cell Mol Life Sci 6(6):201–209

Crane MB, Lawrence WJC (1934) The genetics of garden plants. Macmillan and Co., London

Ewing B, Green P (2000) Analysis of expressed sequence tags indicates 35,000 human genes. Nat Genet 25(2):232–234

Fischer E (1902) Synthesis in the purine and sugar group. Nobel lecture. Chemistry. 1901–1921. Elsevier, Amsterdam, 1966

Fischer E (1906) Untersuchungen über aminosäuren, polypeptide und proteine. Ber Dtsch Chem Ges 39(1):530–610

Fruton JS (1985) Contrasts in scientific style. Emil Fischer and Franz Hofmeister: their research groups and their theory of protein structure. Proc Am Philos Soc 129:313–370

Griffith F (1928) The significance of pneumococcal types. J Hygiene 27(02):113–159

Hershey AD, Chase M (1952) Independent functions of viral protein and nucleic acid in growth of bacteriophage. J Gen Physiol 36(1):39–56

Hofmeister F (1902) Über Bau und Gruppierung der Eiweisskörper. Ergebnisse der Physiologie 1 (1):759–802

Levene PA (1919) The structure of yeast nucleic acid. IV. Ammonia hydrolysis. J Biol Chem 40:415–424

Liang F, Holt I, Pertea G, Karamycheva S, Salzberg SL, Quackenbush J (2000) Gene index analysis of the human genome estimates approximately 120,000 genes. Nat Genet 25 (2):239–240

McCarty M (1985) The transforming principle. W.W. Norton, New York

Vauquelin LN, Robiquet PJ (1806) The discovery of a new plant principle in Asparagus sativus. Ann Chim (Paris) 57(2):1 [Vauquelin LN, Robiquet PJ (1806) La découverte d'un nouveau principe végétal dans le suc des asperges. Ann Chim 57:88–93

Wollaston WH (1810) On cystic oxide, a new species of urinary calculus. Philos Trans R Soc Lond 100:223–230

Difficulties of Chromosome Handling and Access to Material

5

Almost as soon as chromosomes were shown to be inherently important in both plants and animals, questions were raised as to their position in human heredity, but there is a problem with human chromosomes. There are many problems of handling chromosomes because of their scale, but with human chromosomes, this has all to do with the chromosomal carrier rather than the chromosomes themselves. People are not good experimental organisms when it comes to production of cells suitable for cytogenetic analysis. The problem is one of gaining access to material in which active cell division is taking place. This generally requires some sort of gonad biopsy which is not usually given voluntarily. The ways that are found to get around this were surprising if not entirely acceptable by modern standards, the background to which I will detail later.

The search for the human chromosome number really took off after it was realised that chromosomes were pivotal to Mendelian inheritance. This was even before DNA was recognised as the material of inheritance; the two strands of research into inheritance, the chemical and the physical, were still moving on their separate trajectories. Some discussion about chromosome numbers had started to take place in the nineteenth century. Once it was realised that chromosomes were apparently consistent in number within a species and most probably within the tissues and organs of an individual, speculation as to what the number is might be started. It is important to realise that while chromosomes were being looked at in detail, since they were functionally anonymous, there was no implicit assumption that they were consistent. Some considered it possible that their number was tissue specific, as if they had a mechanical function within the cell forming a specific tissue.

One of the earliest suggestions of a chromosome number in humans came from Hanesmann in 1891 who had looked at what he described as normal material and reported cells of 18, 24 and more than 40 chromosomes (Hansemann 1891). Presumably because of the range he found, there was no definitive statement as to how many he actually thought were present in each cell, although he did suggest that the number was probably 24. The range was an inevitable consequence of the

© Springer International Publishing Switzerland 2016
W.J. Wall, *The Search for Human Chromosomes*,
DOI 10.1007/978-3-319-26336-6_5

preparatory techniques which had to be employed. It was going to be more than half a century later that whole cell preparations could be used, until then it was all about embedding tissue in wax blocks and cutting sections. This is always going to cause problems for researchers in this field. With this uncertainty inherent in the techniques, Hanesmann reported 'Die Zahl sicher höher als 24 sei' (The number is certainly higher than 24).

Interestingly there were many attempts to answer the question of what our diploid number is in the 20 years following Hanesmann, but these were primarily associated with trying to determine what mechanism was in action to determine human sex. One of the most interesting attempts in this period, again using sectioned material, was carried out by Winiwarter who reported in 1912 and later that the chromosome number was 47 in men and 48 in women, with the denial of the existence of a Y chromosome (Von Winiwarter 1912). This again was more a serious attempt at devising a model for sex determination than the number of chromosomes itself. H. Von Winiwarter was a remarkable figure in many ways; his primary sphere of work was as a gynaecologist and surgeon, and there is no doubt that this was his primary source of material, but he was also something of an aesthete, not only lecturing on Japanese art but also producing a book about the works of two Ukiyo-e printmakers. This was *Kiyonaga et Choki, Illustrateurs de Livres*, published in 1924.

Since the material which Winiwarter used was almost a by-product of his gynaecological studies, he was not always clear in his publications as to what he had made use of. On at least one occasion, however, we do know where his material came from. He describes discovering what he called an 'internal hermaphrodite' following a hernia operation. Such a finding could explain his description of 47 chromosomes, always remembering that Winiwarter was using sectioned tissue, a notoriously difficult material to handle. There is another and quite possible explanation in that this internal hermaphrodite, which was not detailed in structure, might have been a hydatidiform mole. This is a form of placental tumour, although these do not usually have an abnormal chromosomal constitution. It was for many years a routine to test by way of buccal smears any female child being considered for surgical intervention for an inguinal hernia. The reason for this was to clarify as to whether there was any possibility that the surgeon would find internal gonad material causing the hernia. The buccal smear would be simultaneously fixed and stained in acetic orcein to show up the presence of Barr bodies, the condensed X chromosome attached to the nuclear membrane.

The work of Winiwarter and many of the later geneticists pursuing the human chromosome number were hampered by a requirement for a lot of time to be spent on each microscope slide. This is still the case, but at the beginning of the twentieth century, the use of cameras attached to microscopes was unknown. The only way of consistently producing an illustration of the view down a microscope was via a camera lucida. This device was a system whereby the image as viewed in the object pane of the microscope was projected onto a piece of paper next to the observer, who could then make what was in effect a tracing of the image. This may seem a rather old fashioned system, but even now repeated observation of microscopic material and drawing of images can result in a far greater understanding of pathology than simply looking at photographs. This is in part because a drawing can encompass details from several planes of focus simultaneously. Although autofocus microscopes can be used to take multiple images in a section and sophisticated software then used to reconstruct a composite image, these systems are best employed in specialist investigations where specific aspects of the sections are being investigated. This is the technology which is used in specialist diagnostic services. If the investigator is trying to extract as much information from sections on a slide as is possible, drawing may still be the best option because it may not be known what a significant feature is and what is not, until many slides have been looked at and images produced.

Attempts to replace drawing, which is a learnt skill, started with low-resolution photographic images. These early photographs had magnifications below about x20 and were made by William Henry Fox Talbot in 1839, but these were of little use for any practical purposes. The images were not clear and the magnification too low. It was later, in 1852, that F. Meyer of Frankfurt produced the first piece of apparatus which had any likeness to professional photomicrographic equipment. One of the major problems with this sort of equipment was vibrations. This is a logical consequence of size and magnification; an apparently tiny movement will be magnified as well as the image. A simple way of overcoming this was to effectively lay the device onto its side so that support was available along its entire length. These horizontal systems were widely used until both Zeiss and Leitz developed vertical systems which worked well. Moving towards a vertical microscope and camera arrangement allowed manufacturers to produce single pieces of equipment of dual use, rather than having to make two different pieces of equipment, one to view and one to photograph, which was how the situation was originally approached. Even so it was not until the 1930s that cameras and microscopes combined became commonly available and 40 years later before every microscope routinely used in cytogenetics would have a camera attached.

It was in 1923 that Theophilus Painter (1889–1969) published a very influential paper in the *Journal of Experimental Zoology* (Painter 1923). This was a considered appraisal of the known data available regarding the human chromosome number, which included his own data. This was not the first of his publications on the subject, but it was certainly the most important. Previously, in 1921 and 1922, he had ventured on possible human diploid numbers of 46–48 (Painter 1921, 1922). In 1922 he had decided it was 48, with a sex determining system of XY. These were all based on observations of sectioned tissue. Prior to this the general feeling was that humans had 24 chromosomes, with various suggestions for a range of different methods of sex determination. These various mechanisms mirrored the range of different ways in which sex determination was seen in other plants and animals. Painter reported that with the exception of Winiwarter, all the other studies of human chromosomes had been undertaken on spermatogenesis, using 'stale tissue'. This is a phrase which requires an explanation. Gaining access to human material was never going to be easy, and for most researchers in this field, whatever was available had to be used. It was normal for interested physicians to have access to

Fig. 5.1 (**a**) Illustration of the analysis of T. H. Painter from 1923. This was carried out using a Bausch and Lomb microscope and drawn using a camera lucida. It was a remarkable feat to be able to decipher these chromosomes from squashes and come so close to a correct interpretation of the result. This was originally a fold-out illustration; the twofold lines can just be made out in the image. (**b**) Original explanation for Painter's illustration shown in (**a**). This was Fig. 6 in the original paper of 1923

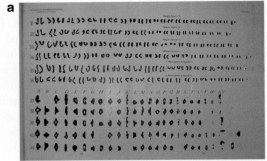

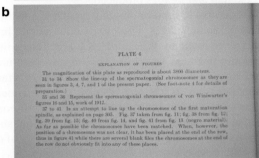

the bodies of executed criminals. This would involve some persuasion and expense on the part of the medical practitioner, but as a source of research material for anatomy and physiology, it was invaluable. So it was often the case that testes would be removed from executed criminals post-mortem for genetic studies. The time between death and access to the corpse was the reason that Painter described the material used as 'stale' (Fig. 5.1a, b).

Theophilus Painter had the advantage of a friendship and collaboration at the University of Texas with Professor Dana Casteel. Casteel was in the Department of Zoology and by chance was acquainted with Dr. T. E. Cook, the physician at the Texas State Insane Asylum. Painter refers to the Texas State Insane Asylum, but it is generally considered that it was completed under the title of the Texas State Lunatic Asylum in 1861, changing its name to the Austin State Hospital in 1925. As the official physician to the Texas Asylum, it was possible for Dr. Cook to pass on material of importance to Prof. Casteel, and from there it went to Painter. One of the reasons that this flow of material was possible was because Cook was engaged with the idea of academic research benefiting medicine, or as it was put, Cook was recorded as being interested in the 'greater medicine'.

The samples which were sent to Painter were not so much testicular biopsies as entire testes. These came from two Negroes and one white male. Painter describes the reason for the castration as 'excessive self abuse coupled with certain phases of insanity'. An hour or two before the operation, the patients were given injections of morphine to quieten them down, and then under local anaesthetic given in the form of novocaine, the operation was undertaken. The patients were recorded as showing no interest in the proceedings and no signs of pain or discomfort except when the vas deferens and accompanying nerves were cut. Once removed the testes were quickly cut into pieces with a razor and dropped into preserving fluid where they were thoroughly teased apart. This teasing apart would have been to try and get the fixative to infiltrate as thoroughly as possible to preserve the cellular morphology. Once fixed, the samples were embedded in wax and sectioned at 4–8 μ, the nucleus being approximately 10 μ in diameter. The results were illustrated using a camera lucida so that the chromosomes, as seen, could be put into pairs, a lot like making a modern ideogram. Don't imagine this was easy. The solid stained chromosomes were morphologically unrecognisable, and having been taken from sections, they were easily counted twice, or if they were overlapping then two chromosomes would be counted as one. The problems of trying to analyse sectioned material remained a major hurdle to understanding the number and displacement of the chromosomes within the nucleus. This is well demonstrated by Painter saying that he found 48 chromosomes in all the cells he illustrated and some with 49 chromosomes. What he was doing was giving a true account of what he saw, hence the variability in his scores.

Having thoroughly analysed his own work, Painter then went back to the paper by Winiwarter and re-analysed it to compare the results. To modern eyes this would be a daunting prospect, but as these were the current best material, it was clear to Painter that there were 48 chromosomes in all. He sized the chromosomes so that there was a clear progression from largest to smallest chromosomes, which he labelled from A (largest) to W (smallest), and then added in X and Y, thereby giving 48 in all.

Interestingly, Painter considered that in his preparations the X and Y chromosomes were joined by a thread, a sort of permanent attachment between the two parts of the XY chromosome unit. This single entity only divides into its two component parts when the cells divide and one meiotic product receives the X part and the other a Y part. When these combine with the other meiotic cell in fertilisation, they form up either an XX unit or XY unit. Painter is also convinced that the larger of the two components of the XY unit is the X chromosomes. In one of the illustrations from the camera lucida, the clarity of his drawing is well shown. He points to two chromosomes apparently attached by a thread with a pair of beads on it. This he says is the XY chromosome showing the thread of attachment. What it really looks like is a finely observed example of the well-known phenomenon of satellite association. The two chromosomes are probably a modern D group chromosome (13, 14 or 15) and a G group chromosome (21 or 22).

Satellite associations are one of the observable phenomenon which clearly indicate that the cell is not a random collection of chromosomes, but a highly organised one. It is not unusual to find G and D group chromosomes very nearly touching, satellites to satellites. In some preparations they appear to be attached to each other. This can cause some confusion because the so-called Robertsonian translocation, or fusion, is a true attachment of D and G group chromosomes at their satellite ends, which has no affect on the carrier but can have severe implications in pregnancy by the transmission of unbalanced meiotic products. Painter published his experimental technique in more detail in 1924, where it was suggested as suitable for students to use instead of relying on chromosomal preparations from invertebrate sources (Painter 1924).

Slightly earlier than the work of Painter, a German called Grosser used cells from the amnion of young human embryos and claimed to have seen 47 in one sample and 48 in another. He also said that he had found chromosome counts ranging from 45 to 54. This was also the work which made use of the material he studied as part of his investigation into the placenta as a method of evaluating human evolution. His work was published in 1909 and 1927 (Grosser 1909, 1927) and was used to back up the claim by Painter that the modal human chromosome number is 48.

The reason that we refer to the modal number is both well known and a simple one, but worth repeating. When cells are taken as a biopsy, there is always a chance that there has been a chromosomal loss in vivo. Even more likely, though, is that once they have been transferred to culture, where the cells can survive without their entire genome, chromosomes may be shed without problems for the culture. During the preparation of slides from cells grown in culture, mechanical damage may occur with the cells losing chromosomes into the fixative solution. For this reason, the number of chromosomes is an average, not the mean but the mode, that is, the most frequently encountered number. The obvious reason for this is there is no sense in taking a mean value after counting several cells as this would result in something meaningless like 45.23 chromosomes.

Painter also noted something which would prove to be true, although frequently contested by later workers. As his key paper of 1921 was based on material from two different racial groups, he could make a definitive statement that the chromosome number was not different between the two races, determined by colour, from which he had received material. Considering the small size of his sample, we might think that this was a rash statement, but we must remember that Painter had looked at large numbers of different species. He developed a certainty of the reliability of chromosome numbers, even though he was working with difficult animal material, so if the number was consistent within an animal species, it should certainly be so within humans, and his small sample supported this idea.

Access to suitable human material was always going to be problematic, but in some cases this was solved through another area of genetics, although in this case most definitely a misuse of science. Although eugenics had a benign start as an attempt to improve the state of humanity, the basic premise upon which it was founded was incorrect. The ideas regarding eugenics were based upon a premise that humanity could be treated, at least in part, in the same way that an animal breeding programme would be. This is, of course, self-evidently incorrect. It was the application of eugenics along these lines with its scientific verisimilitude, perpetrated by politically motivated individuals which rapidly brought the subject into disrepute and caused such enormous turmoil during the twentieth century.

Eugenics can be broadly described as studying heredity with a view to the improvement of the human race. Besides the inevitable criticism that we do not

Fig. 5.2 Francis Galton
c.1860 painted by Octavius
Oakley (1800–1867)

have the ability to decide what constitutes improvement, there is also an obvious link between these ideas and simple parochialism where anyone not of the tribe is inferior. We now tend to think that as a movement eugenics started with Francis Galton (1822–1911) in 1865. Before the word was coined by him in 1883, the ideas were understood from animal breeding but had no umbrella term when applied to people. Or at least that was when he first voiced his ideas of investigating the origins of natural ability in human populations. Galton himself was born in the English midlands into a prosperous family of manufacturers and bankers, and when he graduated from Cambridge in 1844, he was financially independent. His first recognition as a scientist was brought about when he returned from a 2-year expedition to as yet unexplored areas of Africa, and it was substantially for this work that he was elected Fellow of the Royal Society in 1856 (Fig. 5.2).

Galton's scientific investigations ranged widely, but always emphasised the need for quantification of results and a demonstration that results were repeatable and reliable. This insistence on being able to use observation for testing hypotheses and predicting outcomes was epitomised in his production of the first weather map for *The Times* where he coined the term anticyclone. It was, however, the publication in 1859 of *The Origin of Species*, by his cousin Charles Darwin, that inclined him towards the measurement of all things human, even to collecting medical histories of identical twins. One of the measurements which he was particular interested in and made an extensive collection of was the weight of sweet pea seed compared with the seed of the parent. He chose sweet peas because they are self-fertile, and in a complicated experiment involving passing seeds onto his friends who then grew the plants and returned the resulting seeds to him, he plotted the size of the resulting seeds on a two-dimensional plot. At the time that Galton was working, there was no mathematical method available to compare two such apparently different measurements, but he did notice that the association between two measurements could be expressed as an approximate straight line. From this start he eventually produced the technique in 1888 for calculating the correlation coefficient between two variables, although it was in 1896 that Pearson produced a rigorous mathematical treatment of correlation. Galton was hampered when he started applying his methods and analysis to human characteristics; primarily this was because his choice of characters to measure was somewhat nebulous, involving trying to quantify such things as temperament, artistic ability and disease incidence.

What Galton proposed in his early publications about eugenics was to encourage marriages between gifted individuals to increase the mental performance of the country. He based this suggestion on looking at families that had consistently produced notables amongst the nation's hierarchy. We now know that this line of investigation is fraught with problems, but at the time it seemed perfectly reasonable for people in positions of power to assume they were intrinsically superior. It had, after all, been a long tradition for monarchs to claim they ruled by divine right (Galton 1865, 1883). No doubt for Galton he was in this position of believing in his own superiority, and his own wide-ranging extended family which included Darwins and Wedgwoods would have been included in the development of his ideas. What he had not realised was that inheritance is different from breeding. Wealth can be inherited and by implication a good education; these are factors beyond genetics and lead directly to the fallibility of eugenics as a breeding premise since the pool of humanity has greater variance in upbringing than variance in genetic background. Galton had dismissed arguments such as these in favour of his ideas of breeding and ideas coming out of the nascent science of genetics. There is no doubt that the entire premise of eugenics is unsupportable; it is simply not the same as breeding livestock. Unfortunately just because an idea is incorrect, it does not stop the misrepresenting of science by way of support for personal and political advancement, which is how eugenics developed. It has always been the case that loud voices trump truth in the popular imagination, even when wrong. It is also sadly true that much of what was pedalled as eugenics was done so for entirely the wrong reasons. So we have two shabby notions playing out as truth: the first, that eugenics has something to add to humanity, and the second, that you can justify anything in pursuit of power.

This attitude to producing a better human was compounded later in the nineteenth century with a very short book of only 39 pages, titled *The Rapid Multiplication of the Unfit* written by Victoria Woodhull and published in 1891. The change which occurred here was a move away from just encouraging the fit to produce more children towards encouraging those perceived as unfit to breed less. This was

Fig. 5.3 The plaque commemorating Martins Bank in Liverpool

without doubt an extreme attitude and not one which was universally accepted. Victoria Woodhull was quite a character having been born in 1838; she married, divorced and created many scandals in the USA before moving to the UK, rather under a cloud. While in the USA she became a presidential candidate and opened with her sister the first female brokerage on Wall St. She moved to the UK in 1877 where her life became quieter and she married a banker, John Biddulph Martin. Martin's bank was one of the High Street banks, the only one with its headquarters outside London, until it was subsumed by Barclays in 1969. Although *The Rapid Multiplication of the Unfit* was an ill-conceived and minor work on eugenics, it does serve to demonstrate that such ideas were at that time both acceptable and commonplace. Woodhull had previously published a pamphlet, of only 31 pages, on this same subject in 1888, a very early example of eugenics (Fig. 5.3).

It was in the USA that some of the more extreme attitudes towards eugenics emerged, fuelled by a perceived swamping of the original Anglo-Saxon stock by large numbers of immigrants from southern and eastern Europe. This was illustrated by the strangely titled American Breeders Association. Started in 1906, this was broadly a eugenic society and courted some ill feelings amongst geneticists of the time by having a journal which was sometimes 'reckless' as Thomas Hunt Morgan put it. The American Breeders Association was started by Charles Davenport who was greatly influenced by the work of Galton on biometry, although he did not himself invent the technique. The originator of quantified biometry was Alphonse Bertillon who used a complicated set of measurements that had to be carried out by highly trained technicians, the idea being to measure an individual so that they could be reliably identified in the future. Galton was quite critical of the techniques used as he saw that many of the measurements were the same, just carried out in different ways. One enduring aspect that Bertillon did manage was the introduction of what we would now call the mugshot (Fig. 5.4).

Alexander Graham Bell believed in a positive approach to eugenics and the encouragement of the fit. It was in his capacity as a proponent of positive eugenics that he was the honorary president of the Second International Eugenics Congress of 1921. The first congress had been held in London in 1912. Unfortunately the

Fig. 5.4 Alphonse Bertillon

negative aspects of eugenics gained a certain traction amongst the political groups and made considerable changes to social ideas regarding the less fortunate. Somewhat surprisingly one of the countries which seemed quite keen to embrace the ideas of discouraging the unfit to breed was the USA. The gradual change in attitude towards the people referred to as undesirable or feeble-minded was fuelled by people in positions of power believing in their innate superiority. Unfortunately they did not recognise their own hubris. By 1914 about 30 states in the USA had produced laws which either directly or indirectly controlled the legal marriage of the insane or feeble-minded.

In the UK, for various reasons, the control of proliferation of those described as mentally deficient was pursued inadvertently by having sexually segregated institutions. It is also interesting that while eugenics started as a subject in the UK, it flourished only briefly with the establishment of the Eugenics Education Society in 1907 at the instigation of Sibyl Gotto, a social reformer, and Francis Galton. They published a journal called *The Eugenics Review*. Over many years with the decline in the public acceptance of eugenics, the Society transformed itself into the Galton Institute. It was around the time of the First World War that genetics as a subject in its own right began to exert itself and many of the new breed of geneticists started to distance themselves from the older and more draconian ideas of eugenics.

In the USA the route that was taken to try and control the development of the population was via sterilisation. Although extreme voices had occasionally been heard advocating euthanasia, this had never been seriously contemplated. The first state law for sterilisation was enacted in Indiana in 1907, and over the next 10 years, 15 more states passed similar statutes. Gradually more and more states took on the mantle of the eugenic community with either voluntary or in some cases compulsory sterilisation. Strangely, the reason cited for these statutes, which was mostly based on hearsay evidence, was to curtail excessive sexual activity by the process of sterilisation. This was not going to work in quite the way expected as tube ligation or vasectomies were the normal method used. This would reduce pregnancy but not the hormonal basis of sexual activity. These methods were employed as speedy and economical, but were not specifically stipulated. It was for this reason that it was possible to use castration as a method of sterilisation which would render material available for cytogenetic study. Charles Davenport, of the American Breeders Association, was of the opinion that if sterilisation was to be undertaken, then it should be via castration so as to render the individual docile and without sexual appetites.

Searching for the human chromosome number using somatic material was never going to be an easy option since cell division is not a common aspect of stable and undamaged tissue. On the other hand, gonadal material, testes or ovaries, would normally be expected to contain actively dividing cells. These active cell divisions would be a mixture of mitosis and meiosis, that is, reduction divisions moving towards haploid cells, but the nuclear divisions would have visible chromosomes present, and in the first divisions these would be diploid. The important part is that as long as active cell division is taking place, it can be expected that chromosomes will be visible.

Having access to this actively dividing material is only part of the story of course. Most of the work was being done on serial sections. It is true that squashes of fresh material were possible, but it was very difficult. The problem is that unlike plant material where the cells are sufficiently rigid to withstand physical disruption of the tissue, and individual cells can be separated out by mechanical action, animal tissue is quite different.

Plants have a cell wall which is primarily cellulose and gives not just the cell but the plant as a whole a certain amount of structural integrity. Discounting lignified wood, which is intrinsically rigid, plant material maintains its integrity by using water pressure to hold the structures rigid. In animals the rigidity of structure is conferred by the skeleton, the cells having little strength to maintain their own shape. So if a small piece of plant material, such as a growing root tip, is put on a microscope slide, coverslipped and then tapped lightly, the cells will separate while maintaining their overall shape and contents due to the strength of the cell wall. If these are now squashed with gentle pressure, the cells collapse and the chromosomes can be clearly seen. With animal material the problems start with trying to separate the cells so that individuals can then be stained and squashed. The cells tend to burst, distributing chromosomes about the remains of the cell.

The alternative of sectioning actively dividing materials also carries problems, mainly that it is a very time-consuming process even to sort out a single cell. As long as the cells are sectioned at a fine enough level, it should be possible to gain some sort of an idea of where chromosomes start and stop within the nucleus. But this is not easy. When the cytologist Tao-Chiuh Hsu managed to view a slide that Theophilus Painter had used in his research, he was amazed that anything could be discerned from the ball of wool that was the overlaid chromosomes. It would take a skilled observer a long time to be able to gain anything meaningful from these sorts of preparations. It is, under these circumstances, even more astonishing to consider that he came anywhere near to the correct human chromosome number. Until the middle of the twentieth century, researchers were looking at mechanical sections, cut as finely as the technology of the day permitted. By the turn of the twenty-first century, computer-aided equipment was available to take optical sections through material, a method routinely used in looking at gene amplification, but even with all these aids, it is not an easy process. To try and track through a whole nucleus, following individual chromosomes, one section at a time would try the patience of the most sanguine individual.

It was going to take a change in technique before a major breakthrough was going to be possible. Like so many stages in chromosome hunting, the process is technology driven because the targets of investigation are so difficult to handle and manipulate. It was not far in the future before changes to the way material was handled would make a significant change to human genetics.

References

Galton F (1865) Hereditary talent and character. Macmillan's Mag 12(157–166):318–327
Galton F (1883) Inquiries into the human faculty and its development. JM Dent and Company, London
Grosser O (1909) Vergleichende Anatomie und Entwicklungsgeschichte der Eihäute und der Placenta. Braumüller, Vienna
Grosser O (1927) Frühentwicklung, Eihautbildung und Placentation des Menschen und der Säugetiere, vol 5. JF Bergmann, München
Hansemann D (1891) Ueber pathologische Mitosen. Virchows Arch 123(2):356–370
Painter TS (1921) The Y-chromosome in mammals. Science 53:503–504
Painter TS (1922) The sex chromosomes of the monkey. Science 56:286–287
Painter TS (1923) Studies in mammalian spermatogenesis. II. The spermatogenesis of man. J Exp Zool 37(3):291–336
Painter TS (1924) A technique for the study of mammalian chromosomes. Anat Rec 27(2):77–86
Von Winiwarter HE (1912) tudes sur la spermatogene'se humaine: I. Cellule de sertoli: II. He´te´ rochromome et mitoses de l'epitheleum seminal. Arch Biol 27:91–189
Woodhull V (1888) Stirpiculture: or, the scientific propagation of the human race. London, England. Private publication.
Woodhull V (1891) The rapid multiplication of the unfit. The Women's Anthropological Society of America, New York

The Implications of DNA Structure

6

There are many different aspects to genetics that are self-evident, but it is sometimes not fully appreciated how very complicated it is. The complexity comes not from the individual aspects but from the level of interactions which take place within cells and organisms. This makes understanding the whole picture a difficult task. This inevitably leads us to look at individual aspects of genetics, however misleading the image of simplicity that this generates. Consequently we can become isolationist in looking at a single aspect of inheritance, whether these are visible phenotypic Mendelian characters, of which there are surprisingly few, or the chemical structure of DNA. Most visible phenotypic characters, such as height and hair colour, are either caused by interactions of multiple genes or interactions between the environment and genes. The more important question is "how do all these disparate ideas fit together?" It is all very well that saying something is dominant or recessive, but what makes it one or the other if they are just chemical structures working in isolation? The short answer is, of course, that they don't work in isolation. But it is important to find out the individual parts of the conundrum before the whole picture can be put together. Part of this fitting together is to ask the questions which can be answered, so just as cytogeneticists were looking at chromosomes and their numbers, biochemists were investigating the basic chemistry of the genetic material, and after it had been demonstrated to be this strange and apparently simple molecule DNA, how did it do it?

It was always hoped in cytogenetics that knowing the structure of DNA would give greater insights into chromosome structure and function than it actually did. The problem is that chromosomes are more than DNA; they have a complicated structure of their own with protein being a key controlling agent in both the structure and the function, but not the information content. They sit in the nucleus in a regular arrangement, not randomly. This was going to become obvious to routine clinical cytogeneticists almost as soon as services started as constantly observing metaphase plates would indicate a tendency for certain chromosomes to associate together or for certain chromosomes seemingly to be positioned within the metaphase. This is not to suggest that a chromosome could be identified by its position, but the astute observer would note that a certain chromosome, once identified, was in the same relative position, again. Almost like a subconscious statistical analysis of endless cells, the data pointed to non-random positions within the nucleus. Chromosomes are certainly not just flopping about, their structure and position being pivotal to the correct function of the genes. So it was hoped that the ideal starting place in building up this picture would be DNA structure. The number of chromosomes was only to be found after the structure of DNA had been elucidated, almost as though it did not matter. In fact the structure of DNA and the number and importance of chromosomes were two questions which at the time were seen as completely unrelated. It is quite likely that had you asked the biochemists investigating DNA what the significance of chromosomes were, they would have replied 'very little'. During the 1940s and 1950s, and when the first partial human genome sequence was worked out, it was jumped on as answering all questions regarding structure and function of humanity. This was to be a short-sighted and incorrect assumption based on a lack of understanding of just how complex biological systems can be. It is however just these simple questions, like the structure of DNA, with straightforward and clear-cut answers which catch the imagination of the press and general public. Such things can be lauded and explained in relatively short paragraphs. This is not to reduce the importance of these discoveries; it is to show that even though the chemistry is simple, it fits in to a picture so vast and majestic that it quite takes the breath away to contemplate the implications of it all.

Once it became widely accepted that DNA was the agent of heredity, there were two major groups of questions that arose. These were broadly about either how DNA somehow makes the organism or how DNA maintains its integrity from generation to generation. Both of these were, correctly at the time, seen as broadly chemical investigations. It was correctly seen as chemistry because you could extract DNA and see the material of heredity in front of you, without associated proteins or any other trappings of biology. It was going to be a long time before it became apparent that biology was going to exert a very powerful system of controls and feedback systems that ensures the cell does not inappropriately express genes, regardless of the sequence present.

Our red blood cells have no nucleus, a consequence of which is that red blood cells have no DNA. Where DNA is present, it should be remembered that it is always the same DNA, regardless of whether the cell is in your eye or your intestine. The differences between tissues are a result of the control of gene activity. Before big questions regarding how these controls are exerted could be asked, it was essential that basic knowledge of the chemistry of DNA was worked out.

The first aspect of DNA which was a puzzle was the simplicity of the molecule; simple chemical analysis had demonstrated this, so how did four bases do so much? They were, after all, chemically both simple and only came in four different varieties. In previous years, prior to the work of Avery et al. (1944), there had been the belief, sometimes spoken, sometimes tacit, that to create a complex organism a complex chemical, such as proteins, would be required. What had been forgotten, or more likely not realised outside the confines of mathematics departments, was that simple things can result in great complexity. In more recent times, this idea has become more commonplace as we all function with the help of various types of computers, and we all know that the digital world is ruled by the simplest code imaginable, based on 1 and 0. But on/off codes are not really suitable for biological system where the unit of a code should be of fixed length. It is a simple observation that as codes become longer, they become more prone to errors, so short is best.

It was already known that genes are coded for polypeptide chains—and nothing else. The string of amino acids which results from transcription and translation of a DNA strand will always start off as a linear product. It is only afterwards, when it is folded and modified in all the different ways it takes to make a protein, that it becomes an active component of the cell. This linearity of protein and DNA, made from their respective building blocks, was in itself an interesting discovery because it hinted directly that the two may be colinear. It was felt that if a protein was going to be a linear sequence of amino acids and the nucleotides were linear sequences, there was going to be a code involved, some mechanism to turn one message into another. It turned out to be easier to determine the structure of the double helix than the next stage of what it does and how it does it. The difference is a philosophical one. Determining the basic three-dimensional structure was going to be very dependent upon the development of technology, but the determining of the nature of the code was going to be a thought process, taking as much data as it was possible to find and setting up a hypothesis that could be tested.

While the determination of the structure of DNA was an activity based on sound analytical techniques, the correct result could only be tested indirectly by setting up a hypothesis regarding the replication and transcription of the structure. This is in contrast to devising a code that can be immediately falsified by testing a hypothesis. According to the current thinking of the time, as put forward by Karl Popper (Popper 1934, translated 1982) science should always be made up of testable hypotheses. If this were not the case, then we would be simply observing and recording, nothing more than a sort of natural history. Although the ideas may start as collections of observations, as in the case of Mendel, they quickly become predictive, generating testable hypotheses. It is interesting that the structure of DNA was elucidated and yet did not tell us anything about how it did what it did, although it was guessed that all the information was there to be extracted. It was the very robust nature of the analysis which made it so easily acceptable. Unlike the chromosome number which for the previous 50 years had been open to interpretation and debate because it was observational and subjective, the structure of DNA as described by Watson and Crick was an elegant mathematical solution that stood up to any objective scrutiny to which it was put (Watson and Crick 1953a, b; Fig. 6.1).

When it came to the structure of DNA, it was simultaneously seen as a fundamental goal to be achieved and yet, like so many projects pursued throughout history, a piece of scientific information of academic interest only. Few realised the far reaching potential that would be unleashed by this knowledge. What was hoped was that it would greatly increase the understanding of how genetics worked.

Fig. 6.1 Karl Popper, philosopher of science. He formulated the idea that it was not enough for a statement to be counted as scientific by having observational confirmation. It must be possible for a statement to be disproved by the outcome of a conceivable experiment

The idea of controlling it or altering the outcome of gene expression was a long way off.

Several groups were interested in pursuing the goal of working out DNA structure. Starting from the observation that nucleic acid chains could be of considerable length, the question arose as to what the long-distance structure was. Since the molecule was broadly made up of simple chemical units, whether repeated or random, the molecule should have a crystalline structure which could be teased out. This was going to require some very complicated analysis of X-ray diffraction patterns. There seemed to have been a certain amount of scientific rivalry in pursuit of the structure of DNA from groups in London, Cambridge and the USA. The group in America that was the front runner was led by Linus Pauling, who is regarded as one of the most outstanding chemists of the twentieth century. His results moved him towards a model of DNA which was made up of three strands of nucleic acids formed into a helical structure, although this was quickly regarded as being unlikely by the other groups. At King's College in London, the general feeling was that the structure could not be helical at all. The two protagonists here were Rosalind Franklin and Maurice Wilkins, who have been reported as having had a very difficult working relationship. If this had not been the case, they may well have moved the study forward at a far greater speed. The third group, made up of two scientists, were based in Cambridge. The two scientists were Francis Crick and James Watson.

By the time it was possible to look in detail at the three-dimensional structure of DNA, the techniques had already been used for inorganic molecules and for proteins, as demonstrated by Linus Pauling. The technique used X-ray diffraction which gave a great deal of information compressed into a single two-dimensional photographic image. The extraction of the data from the images produced was a long and difficult process, at that time, in the early 1950s, an activity carried out by hand without the aid of computers.

At this stage it should be said that the prevailing system of education meant that it was possible to gain a degree in biology without any more than touching on genetics. Worse probably than that was the situation which developed later when biologists that became geneticists were usually hopeless when it came to chemistry. Consequently while the biologists were talking about genes and how they influenced life, the question of what they were actually made of was skated over because they did not have the knowledge or vocabulary to investigate this basic question. It was left to chemists, more specifically biochemists, to investigate the basic nature of the gene and its chemical makeup, but again there was a discontinuity of skill. Chemistry could not of itself say anything about the manner in which the three-dimensional structure of the gene functioned to produce the cells and tissues which make up an organism as complicated as a human. It was also realised that it was the three-dimensional structure which would influence the form and structure of chromosomes. To determine the three-dimensional structure required more than just chemistry; it required physics. This was the first time that multidisciplinary science was brought to bear upon a single question with such spectacular results, opening up a whole new subject to detailed investigation from the atomic level right up to multicellular organisms.

Into this story came an interesting group of diverse individuals with different scientific backgrounds. Some of the most productive of these were originally physicists and chemists moving into biology. One of the most prominent chemists was Rosalind Franklin at King's, while amongst the physicists were Maurice Wilkins, also at King's College, London, and Francis Crick in Cambridge. This did signal a move towards a rather more non-biological interpretation of life, going back to basic structures as a stepping stone towards the complete explanation of higher functions, disregarding the immense complexity of interactions between organic molecules and giant biological polymers.

This is not to induce any concept of teleology or to belittle the need to look at the minutiae of mechanisms, but it did lead on to jumping to conclusions which are not necessarily correct when interpreting biological systems. By 1951 the Cavendish Laboratory in Cambridge already housed Francis Crick and James Watson, whose first degree was zoology, under the directorship of Lawrence Bragg, himself a Nobel Laureate in the field of X-ray diffraction. The first attempts to match up the X-ray diffraction patterns of DNA they were aware of with a model that explained them incorporated a helix with the sugar phosphate backbone on the inside. This was recognised as having some aspects which were going to be difficult to reconcile, but it was thought that it might be possible. The next idea was rather more complicated than reality, but given the X-ray data Watson and Crick had access to, this is not so surprising. They had a different way of working which broadly required the making of three-dimensional models that satisfied all the restrictions of the data. So their next idea was a three-strand helix which repeated every 28 Å along the helical axis. When the idea was discussed with the group at King's College, it became apparent that Franklin was not keen on a helical structure as being suitable, but did not offer any alternative ideas which could be tried against the data (Fig. 6.2).

A step forward came when it was realised that the base equivalents that had been shown between adenine and thymine, and guanine and cytosine, also reflected the

Fig. 6.2 (**a**) (Sir) William
Henry Bragg. (**b**) (Sir)
William Lawrence Bragg.
The father and son shared the
1915 Nobel Prize for Physics

possibility that they would stick together in those pairings. Although this gave Watson and Crick a useful piece of information, they thought that at one point they had been outpaced by Pauling when they saw a preprint of a paper which he had prepared for publication. Pauling had also opted for a three-strand helix with the phosphate backbone running down the centre. Already with a formidable reputation, anything Pauling produced had to be looked at seriously, but in this case Watson and Crick knew from their own data that his model was not going to work.

There was always going to be a problem with having a three-strand helical structure if the base equivalence was going to hold true. It would mean that there would have to be a very complicated method of balancing the even number of bases within a structure having an odd number of strands. In the case of the Pauling model, there were some other problems which rendered it unlikely to work. When the idea of a two-strand helix was first mooted, it involved a like-with-like base pairing, which would also require some sort of long-term symmetry to explain the base equivalence. Eventually the complete data from the King's group became available to Watson and Crick in Cambridge. This complete data was needed because there were broadly two forms of DNA depending upon the water content, the two resultant but different images, filling in the necessary gaps in the story. With this additional data, it became possible to adopt a double helix, with adenine pairing with thymine and guanine pairing with cytosine as the elegant answer to the structure of DNA. Amongst those working in the field of DNA chemistry, this model was instantly accepted, but being based on the almost impenetrable data of X-ray diffraction studies, at least to biologists, a more direct demonstration was needed to completely convince the wider scientific community that what was postulated was not a theoretical solution without practical application.

Although we now know that the published double-helix model is the true situation, after all, it is even possible to see the structure on electron micrographs. It was contested for some time until replication studies demonstrated it to be correct. It was for this reason that such a momentous discovery was not on the

Fig. 6.3 Such an iconic image as DNA was inevitably going to be a subject of origami

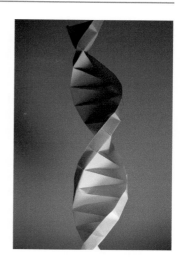

Nobel radar until 1962. Even with the intrinsic ability of the model to predict a manner in which the helix could be accurately self-replicated, experimental data was needed. Just such an experiment was carried out in 1958 (Fig. 6.3).

The division between what was seen as the province of genetics, big-scale cellular activity and phenotypes and the chemical scale of DNA was considerable. Indeed, it was appearing to become greater. There were some questions of chemical structure and function still to be worked out, but the technical gap and the difficulty of understanding more complicated structures such as chromosome structure and function were lagging behind. When Watson and Crick produced their seminal work, it was still some years before the human chromosome number would be determined. This mismatch between the subjects was going to prove very difficult to resolve. There is no doubt that while chemists might be seen as the instrument makers of the scientific orchestra and the molecular biologists as the musicians, it is the geneticist who is the conductor. Just as in the orchestra, it is the conductor who is the one that sees the whole picture, appreciating the minutiae of the parts, knowing that interactions build to a picture far larger than can be seen by the individual players concentrating on their own small part of the manuscript.

By the time that everyone agreed on the mechanism by which DNA replicated, the human chromosome number had been worked out, but it was knitting together of the stories that would become so important in genetics. Before the human chromosome number was discovered, experiments to determine the workings of DNA were being looked for. In 1954 an American researcher, Matthew Meselson, went to Woods Hole where he met Franklin Stahl, a postdoctoral researcher looking at some of the techniques being developed for the newly emerging science of molecular biology. The working relationship between Meselson and Stahl developed very quickly, and during that summer they discussed at length the sort of experiment which could be used to finally determine how DNA is replicated in vivo.

The problem with models of replication was that they depended upon the Watson and Crick double-helix model being correct. In view of that, it was thought that any experiment which demonstrated the way that DNA replicated would simultaneously either prove or disprove the double-helix model. Onto this stage came Meselson and Stahl. If it was assumed that the double-helix model was correct, then it did lend itself to explaining a method of replicating the entire molecule, a complete and accurate copy being made from the original double helix. The question remained as to precisely how it took place. There were three basic ideas for methods of in vivo reproduction which needed to be tested; these were conservative, semi-conservative or dispersive replication. A quite reasonable assumption was made about the process of DNA replication, that whichever of the three suggested methods was correct, it would be the same throughout tissues and organisms. The reason that it was assumed to be a conserved process was that since DNA was universal across life on this planet, so too would the method of replication. A system which evolved in conjunction with DNA as hereditary material would only have been expected to evolve once as the complexity of the system would make it unlikely to evolve twice. The three different methods of replication which were suggested all had positive points in their favour, so experiments to differentiate them had to be very reliable in their results so that they could not be interpreted the wrong way.

Conservative replication was suggested as one option for DNA replication; in this model the entire helix would be used to generate a completely new copy. After the process was finished, there would be two helices: one would be the original and the other an entirely new copy.

The semi-conservative model of replication, the one which turned out to be correct, was in many ways also the easiest to understand. Since the two strands of the helix are held together by hydrogen bonds, which are intrinsically weak bonds, between the opposite bases, the two strands of the helix can be unzipped without disrupting other, stronger, chemical bonds. From there, each strand becomes a template to reconstruct a complete helix. If this model of replication is correct, then after DNA replication followed by cell division, each daughter cell will be made up of exactly half of the original DNA. This was the method of replication which Watson and Crick proposed.

In the dispersive model of replication, the old helix would be reconstructed in sections, almost as though the DNA is cut into sections and new parts spliced in using the old strand as a template. This would not be impossible, but it would certainly be complicated and would not make use of the elegance of the helix being made up of two complementary parts.

The trick was going to be to design an experiment which would exclude some of the options and hopefully give a conclusive answer. After the initial meeting between Meselson and Stahl at Woods Hole, Stahl got a postdoctoral position at California Institute of Technology. By 1957 they had the results they needed, publishing them in 1958. The experiment they devised was of particular elegance in explanation, but like many such experiments the technical difficulties were considerable, requiring considerable ingenuity to overcome. What Meselson and Stahl did was make use of the difference in mass of the various isotopes of nitrogen.

Nitrogen has an atomic number of 7, that is after all what defines the element, and the commonest form has an atomic mass of 14, the rarer one having an atomic mass of 15. Numerically there is 99.64 % ^{14}N and 0.36 % ^{15}N. Chemically, of course, they are indistinguishable, but they do have different masses which are significant enough to be separable using ultracentrifugation and a density gradient. Growing *Escherichia coli* for several generations in a medium containing only ^{15}N all of the DNA would be fully labelled with ^{15}N. By then transferring the cells to a medium which contained only ^{14}N as a source of nitrogen and waiting long enough for a single round of cell division, it was shown that the DNA was made up of ^{14}N ^{15}N hybrid DNA. After a second round of cell division, the extracted DNA were of two types in equal amounts: strands containing ^{14}N ^{15}N and strands containing ^{14}N ^{14}N. After the third round of cell division, there was the same amount of hybrid DNA present but three times as much DNA containing ^{14}N ^{14}N. This is exactly the result which would be expected if the semi-conservative model was correct. It was by separating the strands using ultracentrifugation on a CsCl gradient which proved to be the lynch pin in generating the results and not only demonstrating the manner in which DNA replicated but by implication underlined by the double-helix structure of DNA (Meselson and Stahl 1958).

One of the pieces of equipment without which this experiment could not have succeeded was the ultracentrifuge. Invented by Theodor Svedberg in Uppsala, Sweden, the principle is simple, but the engineering is complicated. By setting up enormous gravitational forces by spinning, the sedimentation coefficient is determined by the molecular mass and comparative size of the molecule. Although the sedimentation coefficient is measured in units of seconds, the values are so small that it is convenient to use the Svedberg (S) where $1S = 10^{-13}$ s. The sedimentation rate also depends on the solvent, so comparison of one result against another depends on the solvent and temperature as well. Theodor Svedberg, who spent his whole career at Uppsala, was awarded a Nobel Prize in chemistry in 1926 (Fig. 6.4).

Fig. 6.4 Theodor Svedberg, inventor of the ultracentrifuge and after whom the Svedberg unit is named

After the classical experiment of Meselson and Stahl had been successfully carried out and it was accepted that DNA replicated semi-conservatively, it was hoped that some sort of chemical explanation of the gene would pop out as a consequence of all the accumulated knowledge about DNA. This was not how it was going to be; although the code for creating a fully functioning protein from DNA has a distinct elegance of simplicity, it took some years to work it out. The story started with the discovery that proteins and DNA were linear arrays of their building blocks. It was impossible not to conclude that in some way, as yet unknown, the linear arrangement of amino acids in a protein was determined by the linearity of the gene. This was underlined by showing that changes in the A protein of *E. coli* was mirrored by a change in the A gene. This colinearity of gene and protein needed a code to explain the link between four bases and 20 amino acids. Or strictly speaking, the code was already there; what was needed was a deciphered explanation of it.

The line of reasoning used to decipher the DNA code started with simple observable facts, such as a single base could not code for an amino acid. That would be woefully inadequate as it would only allow there to be four amino acids. By working up from there using deductive reasoning, the first simple solution which could work would then be the basis for a hypothesis testable by experiment. So, if the code was made up of two bases for each amino acid, then there would still be a restricted range possible, and four bases read in pairs would still only give 16 (4×4) possible permutations. The next possibility would be a code made up of triplets. All of a sudden the permutations become more than adequate. Four bases in groups of three could code for 64 ($4 \times 4 \times 4$) different amino acids (Crick et al. 1961). If this was the case, then there would be a considerable redundancy in the code which would potentially allow for safety against random mutations. This idea was epitomised later when Francis Crick proposed a wobble hypothesis to describe what was then known to be a reduced specificity in the last base of the codon (Crick 1966). There is also a certain amount of redundancy in the first two base positions, which is quite reasonable since with 64 options from triplets of four bases and only 20 amino acids to code for, it is a logical consequence that more than one triplet will code for the same amino acid.

Even though the triplet code gives some resilience to base changes, sometimes it can go awry. The most common mutation in cystic fibrosis is referred to as $\Delta508$, a deleted phenylalanine, the 508th amino acid in the chain. As can be seen in the Fig. 6.5, it is not caused by a triplet being deleted but three bases, from two adjacent triplet codes. It just so happens that taking those three bases out deletes a single amino acid, but closing the gap retains the reading frame of the gene while altering its function. Think of it as removing letters from a sentence; the words are still words, but the sentence no longer works—*Talking again about science* becomes *Talking again out science*. The words work but the sentence does not.

It was Francis Crick and his team that worked on the problem of the code and first showed that it was a triplet that was important. They published their results in 1961 in *Nature*. There are other interesting points regarding the nature of the code, basic questions which needed to be asked and answered to help understand this

Fig. 6.5 The commonest deletion, Δ508, found in cystic fibrosis

ΔF 508

CTT   Deleted causing loss of phenylalanine

506	507	508	509	510	Amino acid number
ATC	AT*C*	***TT*T**	GGT	GTT	DNA code
Ile	Ile	Phe	Gly	Val	Amino acid name

ATC	ATT		GGT	GTT
Ile	Ile		Gly	Val

fundamental aspect of genetics. When it was realised that a triplet code was fundamental to protein production and that it was universal across life, it was necessary to determine if this very same code was an overlapping code or nonoverlapping. If it were overlapping, then reading a gene from a sequence would be extremely difficult, and a single stretch of bases would code for several different proteins and would in effect be several genes in one place. At the very least, a single sequence would have to be read from several different starting points to generate a protein. A very sound argument was put forward against an overlapping code once it became possible to assign base triplets to specific amino acids in a protein. In 1961 Nirenberg and Matthaei managed to start this process which culminated in a complete explanation of the code to amino acid relationship (Nirenberg and Matthaei 1961). Knowing this it became clear that with an overlapping code some combinations of adjacent amino acids would simply not be possible. Evidence of this sort, and constantly looking for the most straightforward explanation of what was already becoming a very complicated system of gene expression, resulted in the nonoverlapping model of transcription being accepted.

The original work was reported two weeks after completion in 1961, the announcement being made at the 5th International Congress of Biochemistry being held in Moscow. Matthaei was not in Moscow at the conference, so the announcement was made by Nirenberg. As is usual at conferences, especially international ones, times are allotted and held to rigidly. As a relatively new postdoctoral researcher, Nirenberg was given 15 min to describe what was a solution to a question that had caused consternation to many larger groups of researchers. Because he had not at that point made his mark, his audience was small, but it was reported to Francis Crick the same day that someone had started solving the amino acid/codon question. He immediately arranged for the programme to be changed and for Nirenberg to be given another spot the next day where he could repeat his results to a much larger audience. The second presentation made an immediate difference and the information was widely and quickly disseminated. In 1968 Nirenberg received a Nobel Prize for the work he had done in understanding the genetic code along with two other researchers, Holley and Khorana. Matthaei was not mentioned in the citation.

Over the decade, the structure of DNA had been elucidated followed by the genetic code. In the same period the human chromosome number had also been determined, but there was still a huge gulf between the two arenas of the micro-scopical and the submicroscopical. It would take a lot to resolve these different perceptions of genetics. In the meantime, there were still questions about how DNA was replicated and how it was packed into chromosomes. Indeed one question which was causing some considerable problems was simply this, why have chromosomes at all? Prokaryotes did not have them, not in terms of the higher-order structures, anyway. They had nucleic acid linkage groups, still a double helix, but comparatively simple compared with the eukaryotes where it soon became apparent that the high-order structure was associated with control mechanisms. Physical control of replication and transcription as well as biochemical control was going to be important in cell and tissue development.

References

Avery OT, MacLeod CM, McCarty M (1944) Studies on the chemical nature of the substance inducing transformation of pneumococcal types induction of transformation by a desoxyri-bonucleic acid fraction isolated from pneumococcus type III. J Exp Med 79(2):137–158

Crick FHC (1966) Codon-anticodon pairing: the wobble hypothesis. J Mol Biol 19(2):548–555

Crick F, Barnett L, Brenner S, Watts-Tobin RJ (1961) General nature of the genetic code for proteins. Nature 192:1227–1232

Meselson M, Stahl FW (1958) The replication of DNA in Escherichia coli. Proc Natl Acad Sci 44 (7):671–682

Nirenberg MW, Matthaei JH (1961) The dependence of cell-free protein synthesis in E. coli upon naturally occurring or synthetic polyribonucleotides. Proc Natl Acad Sci 47(10):1588–1602

Popper K (1934) Reprint (1959) Logic of scientific discovery. Basic Books, New York

Popper KR (1982) Logik der Forschung, vol. 4. JCB Mohr (Paul Siebeck), Tübingen

Watson JD, Crick FH (1953a) Molecular structure of nucleic acids. Nature 171(4356):737–738

Watson JD, Crick FH (1953b) The structure of DNA. In: Cold spring harbor symposia on quantitative biology 18. Cold Spring Harbor Laboratory Press, Cold Spring Harbor, pp 123–131

Tissue Culture and the Cell Cycle: The Answer Is Revealed

It would have been virtually impossible to come to any sort of sound and experimentally verifiable value for the human chromosome number just by using the complicated processes that were routine in the first half of the twentieth century. Besides everything else, there was a technology gap. Trying to gain access to rapidly dividing material, usually gonads, and finding the right cells using techniques that were current at the time was always going to be difficult. The methods, in broad terms, involved taking them through a range of dehydrating alcohols and into a non-polar solvent such as benzene before wax embedding, sectioning and then staining. This was never going to be satisfactory.

That the early searchers after details of chromosomes managed to get so close as they did to the real human chromosome number is very impressive. The number which was generally agreed upon was 48 and therefore inaccurate. This is in itself significant, because it implies a level of imprecision in the results. The counts would range quite widely across the cells seen in each section, so a consensus value from a slide would be arrived at by the individual researcher. This may have been a rounded mean or a modal number. One thing would be certain; it would be as good as possible, studiously searched for. However, since it was an incorrect value, we can assume there were both systematic and statistical errors involved. If the value which was arrived at was 47, then it would be possible to suggest that the material being used was from one of the trisomic conditions we now know to be found in humans from time to time. Since the accepted value for many years was agreed to be 48, this is not an argument that can be made for the results; it was a technical problem. So with this constraint on the reliability of sectioned material as being able to give accurate information, we can be sure that no clinical cytogenetics service was going to be possible, even if it had been considered desirable. It should be mentioned that in some ways the story did come full circle as it is now possible to do some work in diagnostic cytogenetics which can only be carried out on sections.

To create a reliable and accurate number, from which a clinical service could arise, requires reliable and accurate results. This may seem a self-evident truism but is worth stating because it immediately raises the question of how it could be done.

W.J. Wall, *The Search for Human Chromosomes*,
DOI 10.1007/978-3-319-26336-6_7

Fig. 7.1 Wilhelm Roux
(1850–1924)

The answer very definitely resided with looking at the entire chromosome content of whole cells. This is not so easy for many reasons, the first of which is that mammalian cells tend to be attached to each other and teasing them apart damages and kills them. There are two simple sources of nucleated cells which are not attached to each other and can be used: one is white blood cells, but these do not normally divide, and the other is bone marrow, which do divide but are difficult to get hold of. Also, of course, it was not realised at the start of the twentieth century that bone marrow could be a suitable source of material. The answer seemed to be to induce cell division by culturing cells in vitro, not blood cells at this time, but cells from solid tissues, grown in some very elaborate pieces of equipment.

Culture of material from multicellular organisms can be said to have started in a very rudimentary way with Wilhelm Roux in 1885 when he removed the medullary plate from chick embryos and maintained it in warm saline (Roux 1885). This was a step in the right direction, but without a nutrient medium, the cells would inevitably die. In some respects it is a surprise that it took such a relatively long time for the techniques of tissue culture to develop. It is most likely that in general such things would have been seen as aspects of curiosity but not of rigorous scientific endeavour, more technology than science (Fig. 7.1).

Single-cell culture had been practised with great success for millennia, as brewing and in spoiled milk products such as yoghurt and cheese. Here it was that the yeasts were given free rain of the nutritious broth they required to grow and divide, but discovering what was going on in the cultures and the importance of containing spoilage was the province of the nascent microbiological industry. Although carried out on an empirical basis, there was a very large commercial concern in the purity and palatability of their products. So while maintenance of single cells was routine, these were independent cells, not normally associated together. Consequently embryonic organs for short-term use were available to scientists interested in development but were not sustainable in long-term culture. While the goal was the culture of human material, the first recognisable tissue culture was carried out in plants.

In 1902 Gottlieb Haberlandt addressed the German Academy of Science where he described an experiment in which he had single-cell cultures of plant palisade cells in a nutrient salt solution, palisade cells being the support cells lying immediately under the epidermis. Although the cells had stayed alive for a month, there had been no cell division. Interestingly he also suggested that he could see no reason why it should not be possible to reproduce an entire plant from cultured parenchyma cells. This was described in two publications by Haberlandt in 1904 and 1914. Two years later in 1904, Hannig made the first successful attempt at culturing crucifers, now more often referred to as the Brassicaceae, with crucifer remaining an informal term (Hanning 1904).

Of course, growing plant cells in culture poses a different set of problems to growing animal cells, but one thing remains constant and that is the need for sterility. In the case of some plant material, this is relatively easy to ensure, for example, orchid seeds. The normal manner in which they germinate requires a symbiotic fungus, but in culture they can grow on a completely defined nutrient medium, as was demonstrated by Knudson in 1922 (Knudson 1922). So tough is the seed coat that they can be surface sterilised using a hypochlorite solution before being introduced onto the nutrient medium. It goes without saying that this cannot be done with animal material as the sterilising agent would kill the cells you were trying to grow. So for animal cells, sterility has to be ensured from the very instant that the tissue is taken. Other problems associated with the culture of animal cells are based around the growth medium.

Plant growth medium can be very basic, such as a completely defined salt solution which simply provides all the salts and minerals which you would normally expect to find in plant material. It must be said, though, that ill-defined organic ingredients are often added in an almost alchemical way for increased efficacy amongst commercial growers. Additives that have been suggested for better growth include potato, banana and coconut water. In contrast, while it is possible to define the salt content of an animal growth medium, there is no doubt that without the addition of such things as bovine foetal or calf serum, it can be difficult to get any growth or cell division at all. Certainly this particular form of ill-defined additive was routinely used well into the twenty-first century. The first serious attempt to address the problem of a nutrient medium for animal cells was introduced by Ross Granville Harrison in 1907 who was interested in studying developmental problems. Using cells destined to be neurons from frog embryos, he grew them in clotted lymph hanging in a drop (Harrison 1907), a complicated system that was of use for demonstrating the potential of cell culture, but it was not going to be a significant feature of chromosome investigations for a considerable time to come. While animal tissue culture was slowly developing, aspects of botanical genetics which were going to be of inestimable use were also being developed.

Plant geneticists were starting to use a naturally occurring chemical to manipulate the cell cycle, colchicine. This is a plant alkaloid extracted from plants of the genus *Colchicum*, a crocus member of the Liliaceae. Colchicine binds to the tubulin of the mitotic spindle mole for mole, which is why it restricts the movement of chromosomes so that the cell stops at mitosis, unable to separate the chromatids into the two different halves of the nucleus. Colchicine has been used as an active drug for a long time. It is described for use in cases of rheumatism and swelling as far

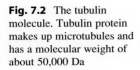
Fig. 7.2 The tubulin
molecule. Tubulin protein
makes up microtubules and
has a molecular weight of
about 50,000 Da

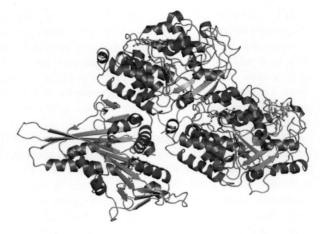

back as 1550 BC in the Ebers Papyrus, although it was the plant rather than
specifically colchicine which was described. The Ebers Papyrus is a text of herbal
knowledge of great age which George Ebers purchased sometime in the winter of
1873–1874 in Luxor, from whom it takes its name. The original of the manuscript
now resides at the University of Leipzig. The first recognised description of using
colchicines in treating gout, rather than rheumatism, gout being the modern medical
use of it, appears in the first century AD when it was described in the five-volume
Materia Medica written by Pedanius Dioscorides. Although he was Greek and
wrote in Greek, he was employed as a medic in the Roman army (Riddle, reprinted
2013). So although it was known that the *Colchicum* bulb had some clinical
activity, it was not known what it was, until in 1820, colchicine was first isolated
in France by a very active pair of chemists, Pelletier and Caventon (1820). It was
known what the active component was and to some extent what it did, but it still
remained unknown as to how it did it (Fig. 7.2).

   Colchicine quickly became a regular and useful part of the armoury of chemicals
used by cell biologists. These were often ill defined in their action and poorly
understood, but they did have reliable and repeatable results which provided a
foundation upon which to develop. In horticulture colchicine was widely used in
plant studies. In 1937 Blakeslee and Amos Avery demonstrated that colchicine
could be used to induce chromosome doubling and suggested that this may be of
value in plant breeding (Blakeslee and Avery 1937a, b). This was confirmed in 1938
when Albert Levan managed to induce huge changes in ploidy levels in plant cells
using root tips of *Allium*, onions, still embedding and sectioning the root tips (Levan
1938). Before these botanical advances, it was being noted that changes could be
made in animal cells with colchicine, although the details remained sketchy.

   Colchicine had been noted to change the blood and bone marrow picture by
Dixon and Malden in 1908, published in the *Journal of Physiology* (Dixon and
Malden 1908), but how the changes were induced was unknown. It was primarily
the work of Lits in 1934 and Dustin also in 1934 that pushed forward interest in
artificial cell cycle manipulation (Dustin 1934; Lits 1934). Both Lits and Dustin

thought from their observations that colchicine stimulated cell division. This is directly contrary to what actually happens but is quite understandable as an interpretation of the snapshot of cell activity which embedded sections afford. Throughout the 1930s, evidence accumulated that colchicine was not a stimulant of cell division but an interrupter of it. Even more precisely, it became apparent that when cell division was interrupted, it was happening at the same point in the cell cycle, mitosis. This was recognised because with the application of colchicine, there was an increasing accumulation of cells at mitosis, and cell division did not seem to progress any further. The experimental method used to investigate the cell cycle and cell division using colchicine was still not based on cell culture, which was still a large technical step away. The normal process was to inject mice either with distilled water (the control) or with varying amounts of colchicine (the experimental) and then record results.

Early investigations involved quite large doses of colchicine, often killing the mice directly, not by any subtle interruption of cell division but by toxicity. These were indeed brutal experiments. What they did clearly indicates that while large doses stopped cell division, it also decreased the blood sugar level and it was this which was considered to be the controlling factor in cell division. Again, this proved to be incorrect, but given the results, it was reasonable at the time to assume that reduction of the available energy supply in the form of sugar would stop the energy-consuming process of cell division. W. S. Bullough in 1949 used a basic and widely copied method to show the effects of colchicine. Mice were injected with colchicine and after varying periods ears were clipped, this being a good source of epidermal cells. The clippings were fixed and sectioned and the mitoses counted. Bullough published on both the affects of colchicine and diet restriction in 1949 (Bullough 1949a, b).

By 1950 it had been confirmed that colchicine itself was the direct cause of cell division stopping at mitosis and not reduced blood sugar, but this work was still being carried out on uncultured material. There was still a significant gap in technology between what was wanted and what was possible which needed to be closed before it was going to be possible for work on cell cycles to be carried out on cultured material. Attempts had been made to artificially culture cells many years before when Ross Harrison had published his work on 'hanging drop' culture in 1907 (Harrison 1907; Harrison et al. 1907). It was only a few years later in 1910 that Montrose Burrows working at the Rockefeller Institute in the laboratory of Alexis Carrel was encouraged to visit Harrison at John Hopkins University and learn the technique. When he returned to the Rockefeller Institute in New York, both he and Carrel started to investigate the range of tissues and species which could be cultured using the hanging drop method (Burrows 1910; Carrel and Burrows 1910; Fig. 7.3).

It had already been established early on that although serum could be used as a growth medium, it was not as good as plasma. The difference being that plasma contains clotting factors such as platelets and ancillary proteins while serum does not. In 1897 Leo Loeb stated that he had cultivated cells outside the body, although the claim was never repeated in print, and when he did publish it in 1902, it was about culturing embryonic guinea pig material in a plasma clot inserted into an adult guinea pig (Loeb 1902). When Carrel and Burrows were reporting their work,

Fig. 7.3 Alexis Carrel,
innovative developer of tissue
culture techniques

Fig. 7.4 Leo Loeb, pictured
about 1915

it had been claimed by Jolly in their previous statements that the cells they had been cultivating in vitro were wrong and that their observations were mistaken. When Burrows returned from Harrison's laboratory, he had already ascertained that it was possible to cultivate cells in vitro from warm-blooded vertebrates, not just from cold-blooded amphibians such as frogs, by using the hanging drop method. This is a technique no longer used as it requires a lot of preparation (Fig. 7.4).

In the hanging drop method, a donor animal has a sample of blood removed which is centrifuged to precipitate the cells and the supernatant plasma separated for use. A drop of the plasma is smeared on to a microscope cover glass and a small piece of finely dissected tissue added. The cover glass is then inverted over a cavity slide which is deep enough for the drop not to touch the bottom. If it is now left, the plasma coagulates, and after some time the cells start to migrate out of the tissue sample and begin dividing. Interestingly, the method of keeping the site of tissue excision and blood taking as clean as possible, if not entirely sterile, that was used involved using olive oil as a local sterilising agent, although this is more of a mechanical barrier than a bactericidal agent.

Both the descriptions and images produced by Carrel and Burrows in their papers published around 1910 are readily recognisable as established cell cultures. These early methods relied on the plasma coagulating to anchor the culture onto the substrate against gravity, whereas in the future gravity would be used to hold the cells or tissue explants onto the growing surface while they physically attached themselves. In Carrel's early experiments, small cultures were inverted, but larger tissue samples, which were too heavy to be supported upside down, were attached to glass plates with plasma and then held vertically in aseptic chambers. Of course, the plasma also supplied nutrients to the growing cells and in the case of embryonic material growth promoters as well.

Using this technique, with a support medium which is a semi-solid clot rather than a liquid, it was important to keep the culture drop thin because if it was too thick, it would allow the cells to grow in three dimensions and make observation difficult. In the published papers, there is always reference to the presence of mitotic figures, this being used as a practical demonstration that the cells are actively growing rather than simply being sustained alive. At this stage there was no specific interest in the group about cell genetics, other than an indicator of cellular activity, so no attempt was made to determine the number of chromosomes present. This is unfortunate because fibroblasts growing on a flat surface are going to be much more amenable to chromosome counting than sectioned material. At this time the cultures would be either fixed and stained on the slides or sectioned and stained; this would be necessary to see the cells and determine their activity. Although it would be possible to see the growing colonies of cells as a cloudy halo extending around the explanted tissue, it would be well into the 1930s before Fritz Zernike started his work which resulted in the phase contrast microscope, for which he received the 1953 Nobel Prize for Physics. With phase contrast, it becomes possible to visualise live cells and be very specific about their morphology and phenotypic description. Although it is not possible to perform a cytogenetic analysis using a phase contrast microscope, it remains an invaluable tool in determining whether a colony has adequate dividing cells to warrant subculturing or processing for final chromosome analysis.

These first steps into in vitro cell culture and the recognition that it was possible to see some cells in metaphase did not significantly help in assessing the human chromosome number. This would come later when the use of colchicine and cell culture came together. By the time that cell culture as a routine technique and colchicine as a metaphase blocking agent were available, the assumption that humans had 48 chromosomes was well established. It is easy to assume that once colchicine was added to the experimental panoply of techniques, it would be a simple matter to count chromosomes from monolayer cultures. This is not so. When cells are growing in a single layer on a flat surface, it is tempting to think of them as being in two dimensions, but of course they are not. The cell might be growing on a two-dimensional surface, but it operates in three dimensions. The mistake of assuming them to be two dimensional comes from an incorrect perception of scale. When a fibroblast rounds up to divide, the chromosomes are in a ball in the middle of the nucleus and the metaphase is in three dimensions. So when this is fixed and stained upon a slide, the cell collapses pushing the content of the nucleus down upon itself and making it extremely difficult to count the crossed over and hidden chromosomes. This would have been very frustrating for researchers who wanted to look at the morphology of human chromosomes. Most of the searchers for human chromosomes would quite likely have previously looked at other species where the chromosomes are fewer in number and much larger. What was needed was some method of separating the chromosomes that were caught at mitosis so that they did not fold over each other and make it virtually impossible for them to be counted and looked at in detail.

Like so many important steps forward in science, the way in which it became routinely possible to produce good-quality metaphase spreads was hit upon purely by chance. Tao-Chiuh Hsu was working in Houston when he came across one of his slides that had particularly clear metaphases, where he saw 'Some beautifully scattered chromosomes' (Hsu 1952). This was an exciting observation, but those cell preparations which he produced afterwards, using what he thought to be the same protocol, were back to normal or as he put it 'resumed their normal miserable appearance'. Hsu was working on cultured cells from skin and spleen taken from a male foetus of 4-month gestation, which were grown on a solid surface. Normally the preparations would be washed in an isotonic salt solution before fixation, but it just happened that on this one occasion, they had been washed in a hypotonic solution before fixation. This was clearly laid out in the addendum to his paper published in 1952. It was attached as an addendum because it had taken a long time to track down the precise reason for this very striking change in chromosome behaviour, which only became obvious when he added water to the salt solution wash. Interestingly, in this paper he still claimed the presence of 48 chromosomes, although he also says that he had trouble finding cells with 48 chromosomes present. He catalogued the chromosomes by length regardless of the centromere position, a pattern which was to become standard across all species. He also said in the addendum that the possibility of hypotonic solutions being useful was being investigated elsewhere, specifically at the Strangeways Research Laboratory in England. His camera lucida illustrations for the paper are clear, but he was obviously influenced in their interpretation by the still current belief that the human chromosome number was 48.

When it became apparent that it was possible to produce cultured cells which could be processed and stained so that they could be analysed clearly, several groups started looking into the possibilities of counting human chromosomes. This culminated in August 1955, when Joe-Hin Tjio, an Indonesian working in Zaragoza in Spain, went to Lund to work with Albert Levan.

Levan had previously spent some time earlier at the Sloan-Kettering Institute in the USA where he had tried out some modifications of the technique introduced by Hsu. When he returned to Lund, he started work with Tjio using tissue from four legally aborted embryos of unknown sex. From the illustrations in the 1956 paper, we can recognise one metaphase which is male and one which is female. The cultures they set up were foetal lung, grown in bovine amniotic fluid. Once established there was a long period in colchicine, after which the cultures were given a hypotonic treatment and then fixed in acetic acid. At this point the technique involved gently squashing the cells on a slide using acetic orcein as simultaneous fixative and stain.

Using acetic acid as the fixative is interesting as it does not cross-link the proteins, which would harden the cells into rigid structures, which is what formaldehyde would do. It is quite apparent that there was a long colchicine treatment from their illustrations, as all the chromosomes show the distinctive contraction associated with prolonged exposure to colchicine.

In their paper Tjio and Levan report with some surprise that 46 chromosomes predominated in all of the cultures from the embryos. They had found lower numbers, but these always seemed associated with damaged cells. On those occasions where they did count 47 or 48 chromosomes, which were very few, they surmised that this was a result of odd chromosomes drifting into the metaphase. They also suggested that the differences could be a reflection of tissue-specific variation. The unexpected finding of a modal number of 46 chromosomes in all of the cultures was considered to be genuine. They did suggest that it was possible, though unlikely, that as they had only looked at embryonic lung fibroblasts, there may be a mechanism which removes two chromosomes from the complete number of 48. Against this they did say that having looked at various tissues from rats where the diploid number was consistent across all tissues, it was reasonable to assume that 46 was a real finding (Tjio and Levan 1956).

By this very important paper, which removed much of the subjective opinion from the interpretation of images of chromosomes, the question of the human chromosome number was convincingly answered. At the end of this article, clearing up the question of the human chromosome number, they make note of another group who had been working on the same investigation but had not made progress because of their own perceived inconsistencies. As they reported:

Dr. Eva Hansen-Melander kindly informed us that during last spring she had studied, in cooperation with Drs. Yngve Melander and Stig Kullander the chromosomes of liver mitoses in aborted human embryos. This study, however, was temporarily discontinued because the workers were unable to find all the 48 human chromosomes in their material.

They go on to say that the most common number they had found was also 46.

In concluding the paper, Tjio and Levan say that although they do not wish to generalise, it would seem that $2n = 46$ is the most likely explanation of their results.

This was quickly taken as a true reflection of being human; the results were repeatable, and unlike sectioned material generally not open to interpretation, the subjectivity that had dogged the century of investigating human chromosomes was over. The process of analysing metaphase plates generated using this technique was much more a simple act of counting, rather than piecing together a picture from serial sections. Even so there were voices raised, if not in dissent, then at least with a questioning edge. One of the areas that generated these questions was whether the chromosome complement was uniform or whether it varied, not so much between tissues, but between racial groups.

In 1958 a paper was published in *Science* by Masuo Kodani who suggested that there was a difference in chromosome numbers between white Americans and Japanese. What Kodani found was that some of the Japanese had 47 and some 48 chromosomes (Kodani 1958). These were suggested to be supernumerary, of unknown origin. However, in this paper he had looked at samples taken from medical resections of testes, not from tissue cultures. The samples were taken and fixed before being prepared as squashes, so the analysis was carried out on meiotic metaphases where the results are not at all clear. Once it became not only routine but also essential to look at somatic cells from cultures, these anomalous results became fewer and fewer until they disappeared.

Soon after the initial work was carried out using tissue culture, it was realised that as a technique it might have some significant clinical applications. Before any sort of service was going to be possible, research into variation and clinical significance needed to set sound foundations of normality against which diagnostics could proceed. For this samples were going to be required. The problem with trying to gain access to human material was one of permission. At this stage a simple blood sample would not be of any use, as there are no spontaneously dividing cells in normal circulating blood.

What was needed was material that could be either induced to divide or was already actively dividing. Into this latter category was bone marrow. It was possible to gain permission for a bone marrow sample to be taken for research, although as it was of purely academic interest as to the result, it does seem rather invasive. Often this material would be taken by persuading the relatives rather than the patient. Although there was no active search going on for a method of resolving the ethical dilemma of taking biopsy material from ostensibly healthy individuals for research purposes, a solution was found. It was important that the question was resolved because in many cases, for example, the severely compromised Down's syndrome cases, the patient was unable to give consent. Under these circumstances, it became necessary to persuade the relatives to give consent on their behalf. Besides all this, it should be remembered that even now, however unlikely, with any invasive technique, there is a risk of complications.

The answer came from what proved to be an unlikely source. Late in the 1940s J. G. Li working with Edwin Osgood at the University of Oregon Medical School in Portland was looking for some sort of additive which they could use to supplement their in vitro culture medium. One of the extracts they tried was from red kidney beans, *Phaseolus vulgaris* (sometimes called *Phaseolus communis*), which they thought had the same growth promoters as chick embryo. Using this extract all they managed to do was agglutinate the erythrocytes, leaving the lymphocytes in suspension. We now know that this is because it binds to N-acetyl-ß-D-galactos-amine on red blood cell surfaces causing them to agglutinate. This was published in 1949, not as a mitogenic agent but as a method for rapid separation of cell populations in blood samples (Li and Osgood 1949). Osgood was already a well established researcher at this time, primarily dealing with abnormal haematology (Osgood 1940).

In the late 1950s, Peter Nowell was working at Pennsylvania University investigating leukaemia. In leukaemia patients, it is not unusual to find spontaneous cell divisions in circulating peripheral blood, but the patients that Nowell was looking at during 1959 were in remission. He was using phytohaemagglutinin (PHA) as a method of separating the cell populations but noticed an abnormally high number of mitoses (Nowell 1960). To find out whether this was a response limited to leukemic cell lines, he went on to test his own blood and found the same result in that there were induced cell divisions. Since his blood was normal, it was extraordinary to find any divisions at all. It was only later that it was shown categorically to be the PHA which was responsible for the reaction. This was not a simple observation, and it was the discovery of the biological activity of an entire group of biochemicals, the lectins. These are now regarded as the best biological response modifiers other than monoclonal antibodies found in nature. When the manuscript was submitted in 1960 to Cancer Research, which eventually published it, one of the reviewers wrote:

It is an interesting observation but of no conceivable significance to science.

It was only a little while after the results were published that T. C. Hsu said that Nowell's work:

was one of the most timely and welcome contributions to human cytogenetics.

The discovery of another mitogenic lectin took place in 1961 and was even more surprising in the way it took place than the discovery of PHA. In 1961 a 3-year-old was admitted to Rhode Island Hospital at Providence with a mysterious and ultimately fatal ailment. At post-mortem it was seen by the Drs. Farnes and Barker that there were cells in the brain like large lymphocytes which appeared to be undergoing cell division. This was not only unexpected but also a very fortuitous observation as it led to the description of the second lectin—pokeweed mitogen (PWM). It turns out that the child had eaten a large quantity of pokeweed berries; this would normally be unlikely, because the berries are very bitter, but according to the parents, the child tended to swallow things without chewing and consequently was not put off by the taste until a fatal dose had been consumed. Pokeweed mitogen can easily be extracted from the pokeweed, *Phytolacca Americana*, and is another method of stimulating division in lymphocytes, although not with the efficiency of PHA, tending to stimulate both B-lymphocytes and T-lymphocytes (Barker et al. 1966).

Another lectin which has laboratory use comes from the jack bean, *Canavalia ensiformis*, and is called concanavalin A. This is used in solid-state immobilisation of some enzymes. There are many other lectins used in biochemical laboratories;

the one thing they all have in common is that they are plant extracts. Lectins are proteins that bind carbohydrates that are in the form of sugars and allow for cell-to-cell interactions without the immunological system being involved. In cytogenetics the most commonly used lectin is PHA. It specifically reacts with T-lymphocytes, giving a uniform population of dividing cells. PWM is one of the few lectins which stimulate T-lymphocytes and B-lymphocytes.

With the introduction of PHA, it became possible for good-quality metaphases to be produced from peripheral blood samples. This was a major step forward because it allowed screening of numerical aberrations of chromosomes to take place. There was a problem, however. Every laboratory had a different way of describing the chromosomes that were being investigated. There was no uniform method of describing chromosomes as individual entities. When it was only possible to analyse a cell from serial sections of actively dividing tissue, such as gonads, or even later when hypotonic treatments were available and the bone marrow was being investigated, it was broadly only by a full description that a chromosome could be identified for further researchers to use.

It was not simply that the process was descriptive; it was also that even when attempts were made to identify specific chromosomes, it was made using an arbitrary system, each laboratory having their own style and vocabulary. For example, when Painter was describing the chromosomes he had identified in the tissue sections from his 1923 paper, he described the chromosomes broadly by size. The centromeres were not obvious structures in his preparations, so it was only size which was used as an identifying character. The method he decided upon was simply to use the designation of A to W, the largest being A, down to the smallest, W. The sex chromosomes were termed X and Y. This was not followed as a convention; many variations upon the idea were tried. Even so, by the time Tjio and Levan published their article, there was still no consistent chromosome catalogue. Consequently, they described how they divided the chromosomes into three groups: M chromosomes (median-submedian centromere; index long arm: short arm 1–1), S chromosomes (subterminal centromere; arm index 2–4) and T chromosomes (nearly terminal centromere; arm index 5 or more). This was broadly the same as Hsu had suggested some years earlier, but there was still no attempt to do more categories of the chromosomes by groups. Lejeune, on the other hand, used a combination of letters (G, M, T, P, C and V) with numbers within each group. With increasing interest in the possibilities of describing chromosomal aberrations, it became obvious that if each group followed their own nomenclature, the eventual outcome would be a chaotic inability for laboratories to exchange data.

What was wanted was a unifying method of describing the human karyotype. By 1960 it was becoming essential for some sort of unifying system, and with the encouragement of Charles Ford, T. T. Puck of the University of Colorado Medical School undertook to organise a conference. It was to be held in Denver and was intentionally kept small. In fact it was broadly restricted to individuals who had published articles on the subject of human karyotypes. This conference managed to gain funding from the American Cancer Society, who recognised the potential of karyotyping for the future of cancer genetics (Denver 1960). The original idea was specifically to sort out the nomenclature as far as it was possible because there were so many different methods being used. As it was thought that there may be a large partisan content, with every group claiming their method was best, there were three geneticists who did not generally work on human material and who were appointed as committee councillors to arbitrate. These were D. G. Catcheside from Birmingham, H. J. Muller from Bloomington and C. Stern from Berkley. Over the 4-day meeting, agreement was reached. This was that the autosomes should be numbered 1–22 and the sex chromosomes retain their designations of X and Y. The autosomes should also be grouped into seven broad classes, A to G, as put forward by Patau. This allowed for an indistinct chromosome in the C group to be described as that without having to be more precise. All the chromosomes would then be ordered by size.

This basic classification, as suggested at the Denver Conference, was accepted, but later it would transpire that there was one exception to cataloguing human chromosomes by size. It was at the time assumed and accepted that chromosome 21 was the additional chromosome in cases of Down's syndrome. This, using the Denver system, implied that it was the largest of the G group chromosomes, 21 and 22. When it became possible to differentiate the chromosomes reliably, however, it was seen that chromosome 21 as described in Down's was smaller than chromosome 22. Although there was a slight attempt to redesignate Down's as trisomy 22, to keep the sizes consistent, this was quickly given up. So it is now recognised that the smallest human chromosome is 21, not 22.

Three years later, in 1963, another conference on human cytogenetic nomenclature was called, this time in London. In the 3 years since the Denver conference in 1960, it had been recognised that some chromosomes have a secondary constriction, also that there is some inherent variability seen in some chromosomes, most specifically the Y chromosomes and what was then thought probably to be chromosome 16 (London Conference on the Normal Human Karyotype 1963). Later, when more accurate methods of individual chromosome identification were available, it was shown to definitely be chromosome 16.

By the time of the next conference in 1966 (Chicago Conference 1966), many new discoveries had been made, so much so that this was to become the most significant decision-making meeting for the nascent area of clinical cytogenetics that had so far taken place. This conference was primarily to consider 'new means for describing normal chromosomes and deviations from the normal complement'. Besides introducing a system to describe additional chromosomes and missing chromosomes, it was acknowledged that structural aberrations needed to be described unambiguously. For this to work, it would be necessary to describe the chromosome above and below the centromere, that is, the short and long arms, separately. This was, in itself, an area of considerable debate. The original suggestion was that the two arms should be designated S for short and L for long; this was objected to on the basis of linguistic chauvinism. Lejeune on the other hand was keen on p for the short arms as a literal use of *petite*. To counter that idea a suggestion was made that the German *kurz* could be used for the short arms. It transpired that this apparently small question of nomenclature was generating a very great deal of heated debate. Eventually it was decided that p and q would not only suffice but be both expressive and significant without having any implications for language. They no longer stand for *petite* and *queue*, short and tail, which would be a little strange, but they stand for the statistical term $p + q = 1$. In words, short arm and long arm together make a whole chromosome.

All of these international chromosome conferences were dealing with chromosomes which were solid stained as there were no banding techniques in use. The stains could be of many different types, the commonest being acetic orcein or giemsa. At the time it was not possible to distinguish all of the human chromosomes from each other with absolute certainty, although some could be. For example, solid-stained B group chromosomes number 4 and 5 are morphologically very similar, while the A group chromosomes, 1, 2 and 3, are easily distinguished in solid-stained preparations on the basis of size and centromere position.

Purely by chance, this situation was going to change; in Stockholm at the Karolinska Institute, Torbjörn Caspersson was looking at intercalating dyes to measure the mass of DNA. The original idea was to use acridine orange, but then he considered the possibility that certain alkylating agents, which have a preference for attaching to the guanine ring, would differentiate G–C-rich areas of metaphase chromosomes. This led him to use quinacrine and quinacrine mustard to stain metaphase chromosomes. Using ultraviolet microscope illumination, he produced the first visible banding pattern on human chromosomes. One of the drawbacks of this system is that the fluorescence fades quite quickly and so requires a photograph to be taken. Being a fluorescence technique, it also means that the image can never be as well defined as an image viewed with transmitted illumination. What it did show was that every chromosome could be individually identified. Caspersson went on to demonstrate that a change in the G–C balance of 5 % will result in a change in fluorescence of 50 %. Reflecting the molecules which were used in this, the patterns were described as Q-banding Caspersson published his results in 1970. While this was a major step forward in unequivocal identification of individual chromosomes, perhaps the most important aspect of this observation of Q-bands was that chromosomes are not uniform; they have a longitudinal asymmetry which implies a spectacular level of packing and control of the chromosome in both structure and function. This work resulted in a flurry of important published papers (Caspersson et al. 1970a, b, c, d) which helped develop ideas of clinical cytogenetics as a useable tool in diagnosis.

While the work of Caspersson had increased sales of fluorescence microscopes, another development took place which gave further impetus to the study of chromosomes themselves. Published in 1970, this was also a spin-off from investigating aspects of chromosome structure. Mary Lou Pardue and Joseph Gall were working at Yale University looking at the new technique of in situ hybridization (Gall and Pardue 1969). This is the method whereby DNA is denatured and then annealed in the presence of specific DNA sequences containing a radioactive isotope which can be registered on a photographic plate. In situ hybridization is still an important tool but is very rarely carried out with radioactive isotopes, these having been replaced with fluorescence probes. What they did was denatured their chromosome preparations using an alkali which produced single-stranded DNA followed by incubation in 0.3 M salt solutions containing highly repetitive A + T sequences. When they had finished their hybridization procedure, they counterstained the preparation with giemsa, an ill-defined stain based on various oxidation states of methylene blue. This produced a very precise staining pattern with dark centromeres and lighter p and q arms. This was originally carried out using mouse chromosomes, but the unexpected result of centromere staining did not go unnoticed. When it was applied to human metaphase chromosomes, it not only demonstrated the presence of centromeres, but it soon became apparent that it was actually demonstrating the presence of heterochromatin. In humans large amounts are found almost exclusively associated with centromeres, except in the case of Y chromosomes where it is found distally on the long arm. The big surprise was that while the Q-bands showed a consistency of pattern, with C-banding there was considerable variation in heterochromatic content on some chromosomes, in some cases it being a massive variation compared with the size of the chromosome.

By now as well as Q-banding, it was possible to produce fluorescent banding with acridine orange, demonstrated by De la Chapelle in 1971 (Chapelle et al. 1971), which gave a pattern exactly opposite that of Q-bands; the Q light bands were dark and the dark Q-bands were light. From this it was inevitable that it would be referred to as R-banding (Verma and Lubs 1975). Both of these techniques suffered from the same problem of low resolution and impermanence of staining as they both fade under the influence of ultraviolet light. While a group at Edinburgh, A. T. Sumner, H. J. Evans and R. A. Buckland were looking at techniques of C-banding human chromosomes, they made an interesting observation (Sumner et al. 1971). The C-bands came out well, but occasionally there was a hint of transverse banding at the same time. It was decided to pursue the transverse bands specifically and they developed a system using acetic acid, saline and giemsa, called the ASG technique which yielded a banding pattern which was an exact copy of the Q-banding pattern, but permanent and visible with transmitted light.

It soon became apparent that quite distinct banding patterns could be generated on human chromosomes using virtually anything that disrupts the association between protein and chromatin. This has been made use of in clinical situations where these G-bands are the staple of cytogenetic analysis. The preferred method of G-banding is by using a trypsin solution as the protein digester; any proteolytic enzyme will do, as will alkaline solutions. The most important aspect is that the eventual banding pattern is always the same, so deviation from the normal tells us of rearrangements. It quickly became apparent that because the structure of eukaryote chromosomes is uniform across species, that is, DNA, histone and non-histone proteins, they were also amenable to the same banding processes. Over the years this has resulted in many different standards for various mammalian species, starting with the International System for Human Cytogenetic Nomenclature (ISCN) (Paris Conference 1971), and being developed for many different groups. The most important aspect to aim for when analysing the chromosomes of a new species is to have the chromosomes in pairs and number them from the largest (number one) to the smallest. As the chromosome number varies widely from species to species, the ideogram will also look quite different (Fig. 7.5a, b).

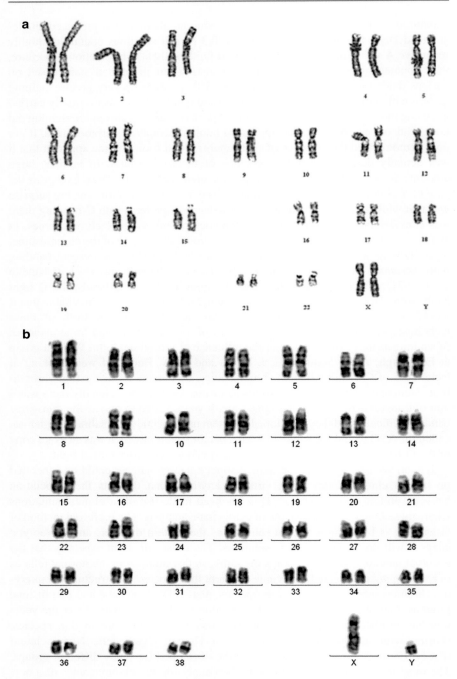

Fig. 7.5 Mammalian chromosomes are remarkably similar in structure and can be g-banded with proteolytic enzymes in the same way. (**a**) Female human chromosomes 46,XX. (**b**) Male canine chromosomes 78,XY

References

Barker BE, Farnes P, LaMarche PH (1966) Peripheral blood plasmacytosis following systemic exposure to Phytolacca americana (pokeweed). Pediatrics 38(3):490–493

Blakeslee AF, Avery AG (1937a) Methods of inducing doubling of chromosomes in plants by treatment with colchicine. J Hered 28(12):393–411

Blakeslee AF, Avery AG (1937b) Methods of inducing chromosome doubling in plants by treatment with colchicine. Science 86:408

Bullough WS (1949a) The action of colchicine in arresting epidermal mitosis. J Exp Biol 26 (3):287–291

Bullough WS (1949b) The effect of a restricted diet on mitotic activity in the mouse. Br J Cancer 3 (2):275

Burrows MT (1910) The cultivation of tissues of the chick-embryo outside the body. JAMA 55 (24):2057–2058

Carrel A, Burrows MT (1910) Cultivation of adult tissues and organs outside of the body. JAMA 55(16):1379–1381

Caspersson T, Zech L, Johansson C, Modest EJ (1970a) Identification of human chromosomes by DNA-binding fluorescent agents. Chromosoma 30(2):215–227

Caspersson T, Zech L, Modest EJ (1970b) Fluorescent labeling of chromosomal DNA: superiority of quinacrine mustard to quinacrine. Science 170(3962):1067

Caspersson T, Zech L, Johansson C (1970c) Differential binding of alkylating fluorochromes in human chromosomes. Exp Cell Res 60(3):315–319

Caspersson T, Zech L, Johansson C (1970d) Analysis of human metaphase chromosome set by aid of DNA-binding fluorescent agents. Exp Cell Res 62(2):490–492

Chapelle A, Schröder J, Selander RK (1971) Repetitious DNA in mammalian chromosomes. Hereditas 69(1):149–153

Chicago Conference (1966) Standardisation in human cytogenetics birth defects: original art. Series II, Number 2. The National Foundation of New York

Denver Conference (1960) Lancet 1:1063

Riddle JM (2013) Dioscorides on pharmacy and medicine (History of science series). University of Texas Press, Austin, TX (Reprint edition 2013)

Dixon WE, Malden W (1908) Colchicine with special reference to its mode of action and effect on bone-marrow. J Physiol 37(1):50–76

Dustin AP (1934) Action de la colchicine sur le sarcome greffe, type Crocker, de la souris. Bull Acad Med Belg 14:487–488

Gall JG, Pardue ML (1969) Molecular hybridization of radioactive DNA to the DNA of cytological preparations. Proc Natl Acad Sci 64(2):600–604

Haberlandt G (1904) Physiologische pflanzenanatomie. W. Engelmann, Leipzig

Haberlandt G (1914) Physiological plant anatomy. Macmillan, London

Hanning E (1904) Uber die kultur von Cruciferne, Embryonen ausenhalb des Embryosachs. Bot Ztg 62:45–80

Harrison RG (1907) Experiments in transplanting limbs and their bearing upon the problems of the development of nerves. J Exp Zool 4(2):239–281

Harrison RG, Greenman MJ, Mall FP, Jackson CM (1907) Observations of the living developing nerve fiber. Anat Rec 1(5):116–128

Hsu TC (1952) Mammalian chromosomes in vitro I. The karyotype of man. J Hered 43:167–172

Knudson L (1922) Non-symbiotic germination of orchid seeds. Bot Gaz 73:1–25

Kodani M (1958) Three chromosome numbers in Whites and Japanese. Science 6:1339–1340

Levan A (1938) The effect of colchicine on root mitoses in Allium. Hereditas 24(4):471–486

Li JG, Osgood EE (1949) A method for the rapid separation of leukocytes and nucleated erythrocytes from blood or marrow with a phytohemagglutinin from red beans (Phaseolus vulgaris). Blood 4(5):670–675

Lits F (1934) Contribution à l'étude des réactions cellulaires provoquées par la colchicine. Compt rend Soc de biol 115(1421):15

Loeb L (1902) Further investigations in transplantation of tumors. J Med Res 8(1):44

London Conference on the Normal Human Karyotype (1963) Cytogenetics 2:264–268

Nowell PC (1960) Phytohemagglutinin: an initiator of mitosis in cultures of normal human leukocytes. Cancer Res 20(4):462–466

Osgood EE (1940) Textbook of laboratory diagnosis. P. Blakiston's Son, Philadelphia, PA

Paris Conference (1971) Standardisation in human cytogenetics birth defects: original art. Series VIII, Number 7. The National Foundation of New York 1972

Pelletier PS, Caventon J (1820) Annales de chimie et de physique 14:69

Roux W (1885) Beitrage zur Morphologie der funktionnellen Anspassung. Arch Anat Physiol Anat Abt 9:120–158

Sumner AT, Evans HJ, Buckland RA (1971) New technique for distinguishing between human chromosomes. Nature 232(27):31–32

Tjio JH, Levan A (1956) The chromosome number of man. Hereditas 42:1–6

Verma RS, Lubs HA (1975) A simple R banding technique. Am J Hum Genet 27:110–117

The Flowering of Clinical Genetics

8

Although genetic diseases, or some of them like haemophilia, had been noted for a long time, their cataloguing and treatment remained nonexistent for centuries. This was understandable because it was unknown what the mechanism of inheritance was. While the working population had a dangerous life and infant mortality was high, it was never going to be easy to follow a family history. For example, cystic fibrosis was not described as a single condition until 1938 (Andersen 1938), and so most probably before then, early mortality would be put down as childhood consumption and later tuberculosis. Even with extensive families, working out relationships and patterns of inheritance is difficult when some members of a family are not talked about.

There is an interesting situation that quickly developed after the introduction of widespread antibiotic use. Prior to the control of infectious disease, this was the major cause of infant mortality. More or less until the middle of the twentieth century, if a child made it beyond adolescence then short of traumatic accident, a long life could be expected; during early life, infections of genetically normal and compromised individuals would be the biggest child health problem. This changed with the introduction of entire families of antibiotics which meant that infections, such as gangrene and septicaemia due to trauma, were no longer feared as inevitably fatal. This was recognised by the Medical Research Council in 1978, as significant enough for specific comment. When they reviewed the situation regarding child health, they opined 'handicaps due to a genetic disorder or congenital malformation, are the major child-health problem today'. By then services had been organised in many countries, but it reflected the changing face of medicine (Medical Research Council 1978a).

The way in which genetic services came about is interesting because the control of what is essentially a very complicated scientific endeavour has inexorably moved from scientific hands to those of the medical profession. This has been with scant regards to the science, which they cannot carry out themselves, so they retain the mystique by being the interlocutor. While for most medical conditions in the past diagnostics was based on the person in front of the physician and the physician were

© Springer International Publishing Switzerland 2016
W.J. Wall, *The Search for Human Chromosomes*,
DOI 10.1007/978-3-319-26336-6_8

only people directly involved in the case, this is not so with genetics. It inevitably involves families, and even the somatic cell mutations that result in tumour induction may be family specific. This makes genetic counselling pivotal to genetic services, but it did take some time for it to become a separate discipline from the strictly medical side of genetic testing.

It is generally considered that clinical services started in the 1950s, before the realisation of the human chromosome number; after all, human genetics did not start with the chromosome. Biochemical genetics were rather more advanced and so they started the ball rolling. Early studies of human genetics were biochemical as the tests were simple extensions of already extant methods, and when new methods for biochemical analysis became available, they were rapidly taken on for clinical applications. This is in contrast to cytogenetics where the clinical curiosity pushed forward the technology directly into clinical applications. The early biochemical tests were generally carried out on urine or blood, the metabolites of interest already being in solution. As the basic technique of carrying out the test was almost always non-specific, a generic assay adapted to a specific requirement, it was well suited to clinical applications. These early biochemical tests were laboratory based, and although sometimes complicated, they yielded a simple result directly related to the physiology of the patient. The only complexity of these sorts of biochemical analyses related to a genetic defect was that interpreting results depended upon knowing something about the family history. In fact even gaining access to the right test required knowing a family history. It was not possible to simply refer an individual for a test because the non-specific nature of the laboratory work made each analysis a bespoke product. Although extended family analysis could track the method of inheritance and probability of future generations being affected, there was no link between what was known within one family and how it related to any other group. The case of haemophilia in the royal families of Europe is one such example. It is possible to track the original mutation backwards to Queen Victoria and then forwards to generate probabilities for descendents being carriers, but more specific testing was until recently unavailable.

There is no doubt that biochemical genetics was essential to the acceptance of a clinical service dedicated to genetic disease, but it tended to remain as a part of pathology for many years. Thalassaemia and sickle cell anaemia had been described during the 1920s, but it was not until the late 1940s that Linus Pauling persuaded Harvey Itano to look at haemoglobin from sickle cell disease patients using electrophoresis. This was the first time that a biochemical difference could be ascribed to a specific genetic condition at the protein level and was the start of molecular disease studies. By the late 1950s slab gel starch electrophoresis allowed the separation of minor components of haemoglobins to be investigated, thus refining the information that was available to the clinician to help with prognosis in such cases. In terms of prenatal studies, haemoglobinopathies posed a very significant problem for families and staff alike.

Early on it became apparent that something could be done to alleviate distress in families with these Mendelian conditions by family tree analysis, but the information when passed on to the patients had the unexpected effect of reducing the birth rate. This was ascribed to the mixed emotions found within the carrier families where even though it would be possible to determine the exact status of a pregnancy if a blood sample from the foetus was available, the potential outcome was too distressing to contemplate. Besides this, the analysis was still not going to be easy because of the different types of haemoglobins produced between conception and birth.

Taking a foetal blood sample was a tremendously skilful job; the risk to the foetus was about 7 %, many times higher than amniocentesis. It also had to be carried out quite late, when there was a sufficiently robust circulation for a sample to be taken (Medical Research Council 1978b). These high-risk Mendelian disorders give about a one in four adverse result; the test is simply to determine which pregnancy is the one in four, at which point there are only two options which have the affected child or a late trimester termination. With this rate of negative outcomes, there was a huge stress put on the parents, but also on the geneticists who were well aware of the results of their analysis.

It was vital to stop these late trimester terminations, and the only possible way of doing that was to find a DNA marker which could be easily analysed, instead of having to take blood for a traditional biochemical analysis. This was luckily found at St. Mary's Hospital Medical School in Bob Williamson's department; all that was required was a very small sample of the chorionic villous, from the growing edge of the placenta (Williamson et al. 1981). Unfortunately the obstetric teams did not really have the equipment to do this sampling. In conjunction with the company, Portex a sampling catheter was developed so that first trimester samples could be taken. The initial results of pregnancies where this technique was used were described in *The Lancet* in December 1982 (Old et al. 1982). Although it did not have any material effect on the progress of diagnosis in April 1983, the Pope said regardless of when, prenatal diagnosis was not acceptable, as though you would take driving lessons from someone who could not drive.

Also in 1982 Zoltan Kazy, with his colleagues Rozovsky and Bakharev, a little ahead of *The Lancet* article of December, published an early demonstration of chorionic villous sampling in the relatively new journal *Prenatal Diagnosis* (Kazy et al. 1982). This did not receive as much attention at the time as might have been expected, partially because the journal had only started publication in 1981 and so was still building a readership and reputation. By gaining access to cv samples taken earlier in pregnancy, cytogenetic analysis also started being carried out at an earlier gestation. Originally the cv tissue was cultured for cytogenetic analysis, and this was quite difficult as it was important to separate the maternal tissue and take cells from areas where divisions were most likely to be found. All of this required both training and skill as well as good microscope work. This time the microscopy was associated with a dissection microscope, needed to see the very small samples in adequate detail to tease the material apart. In 1983 this changed when Simoni in Milan published a technique for getting metaphase preparations from spontane-ously dividing cells within the chorionic villous (Simoni et al. 1983). With the emphasis moving towards diagnosis as early as possible in a pregnancy, techniques were also being developed that allowed earlier amniocentesis, saving a great deal of stress for the parents.

Interestingly there was a publication from Anshan Steelworks, China, in the 1960s where chorionic villous sampling was being carried out for sex selection. This must have been extremely hazardous as they were working without ultrasound (Teaching Hospital of Anshan Iron and Steel Company 1975). Their sexing process was simply a case of looking for the Barr body in the foetal cells, but once it was realised that female embryos were being preferentially terminated on the basis of these results, the practice was stopped. In Russia the situation was slightly different in that since they had to buy their amniotic cell culture medium from overseas, and it did not always work adequately, delays and inadequate storage en route being problematic, so the best alternative was chorionic villous sampling.

It was in 1948 that the American Society of Human Genetics was formally set up, pushed forward by developments in biochemical genetics that were generating results where families, not just individuals, were being involved and inheritance was becoming a major issue. At that time it was recognised that many of the founding members of the society were also members of the American Eugenics Society, which did little to foster good feelings towards clinical genetics. There was also, of course, the recognition that inherited disease was a family burden for which there was little or nothing that could be done; the condition may be diagnosed, but the medical intervention often stopped at that point. In the UK it was not until 1970 that the Clinical Genetics Society was formed, aiming to have both a medical and scientific membership (Christie and Tansey 2003). Formation of these professional societies marks a step forward in acceptance of the discipline as less academic and more practical, although there is always going to be a huge research input into clinical services of this type.

It was after the work of Tjio, Levan and other pioneering cytogeneticists that something interesting happened to clinical genetics. The introduction of a technique which could specifically associate chromosomal changes to phenotypic changes took human genetics out of a backwater and revitalised the subject. It changed clinical genetics from a hanger-on to a discipline akin to anatomy and physiology. This can be seen by the interest which was generated when it was possible to assign a gene to a specific chromosome. This first took place when thymidine kinase was described as being on an E group chromosome, though not until 1971 did Miller assign it specifically to chromosome 17; prior to that date, chromosome identification was insufficient to designate specific chromosomes beyond their alphabetical group (Miller et al. 1971). Once chromosome banding was available, it became easier to be precise in the identification of a specific chromosome.

As far as can be determined, the first cytogenetic tests were carried out in laboratories which were centres of research; there were certainly no laboratories designated for cytogenetics when the requests for tests started, fuelled by publications of experimental results. Interestingly the request for study material came from the research laboratories, while the requests for results from the material came from clinical staff. This hand-in-hand approach between scientific and clinical staff was unusual. Medical tests had often developed out of an ad hoc approach allowing the simple tests to be carried out locally with little fuss in a local laboratory. When cytogenetics started, it was by necessity in laboratories with the necessary equipment; this was a new breed of technology-driven tests. While the nuts and bolts of producing a diagnostic karyotype seems straightforward, in practice it requires a surprising level of technical and scientific ability with very sophisticated equipment, from correct use of sterile materials to microscopy at the limits of the light microscope.

There was a lot of cytogenetic activity developing in Europe. Genetic laboratories were starting to be set up associated with pathology laboratories or as significant laboratories in their own right attached to teaching hospitals. Although the USA had not been slow in developing an interest in chromosomes, it was not uncommon for staff from the UK to travel overseas to assist in setting up a cytogenetic service. Probably the first cytogenetic laboratory specifically for diagnostic purposes was set up around 1960 with the help of Ferguson-Smith and about the same time John Edwards went to Philadelphia where he established their cytogenetic laboratory. Three years later he returned to the UK and started the cytogenetic laboratory in Birmingham with one technician and one microscope. In that particular case, Edwards saw all of the patients before accepting any samples for the laboratory. There was an exception to this which was the buccal smear, these being used as simple tests looking for Barr bodies and are straightforward to carry out on a fixed slide with no cell culture required, the results being available within half an hour of receiving the sample.

With the high technical requirements of cytogenetic services and the relatively low demand, it was always going to be best if laboratories were centralised on a cost basis alone. There was more to it than that; though, if the service was in every hospital, it would make quality control a major and ungovernable issue. Part of the reasoning was that the science behind the tests was intricate, and as said by many clinicians, the result of a genetic test gave a clinical need to see patients who were outside the specialties of other major medical areas. To put it more plainly, genetic diseases involving chromosomes very rarely only affect one organ system and they often echo down the generations, while a broken leg does not.

It was the very close ties between the clinical staff who wanted to know details of genetic conditions and scientific staff who could provide the information in a digested and understandable form which drove the development of clinical genetics as a unified service. This is in contrast to many other disciplines where a physician would simply ask a testing laboratory a particular product and the result would be returned without further comment. While data from genetic testing laboratories fed into clinics, it was the clinical staff who constructed that quintessential genetic tool—the family history. These family trees were often intricate and difficult to produce as, especially in the mid-decades of the twentieth century, older generations did not like intimate details of their family being revealed, especially if they were embarrassed by an institutionalised member of their family. Moreover, of course, it is often impossible to determine, post-mortem, what the cause of death really was, so early death may or may not be due to the genetic lesion being investigated. There was already a great deal of skill associated with the production of risk based on family trees. It was this that was pivotal in deciding the need to set up the genetic clinic at Great Ormond Street Hospital Children in 1946 by John Fraser Roberts.

Throughout this period, there was a pressing need for guidelines in dealing with the patients who came to the clinics. Theirs was not a tale of diagnosis, treatment and cure. Genetic conditions were much more insoluble and needed particularly sensitive handling. The first formalisation of this in print was a book by Sheldon Reed in the USA entitled *Counseling in Medical Genetics* published in 1955. In the UK the first publication dealing with the problems of families with genetic disease did not appear until 1970 with *Genetic Counselling* by Stevenson and Davison (Stevenson et al. 1970).

It was almost inevitable that most of genetic clinics and services would be set up in paediatric units because it was most often a child that was the proband and a paediatrician who was the first to see the child and make a diagnosis. Although the cytogenetic service often stayed within the physical confines of a children's unit, the emphasis shifted slightly away from paediatric diagnosis with the advent of prenatal diagnostic tests. These were a direct attempt to forestall the birth of severely affected individuals by offering therapeutic abortion. Even with the increasing value of genetics being recognised within medicine, it was not possible to capitalise on this as a professional activity in UK medicine until the Department of Health recognised it as a specialty, which only took place in 1980. It was also in 1980 that in the UK the Genetic Nurses and Social Workers Union with a membership of nine nurses and one social worker was started. By 1990 this had increased to a still modest 100 members. Being simultaneously inventive and new to the process of setting up a society, it was realised that some form of written constitution was needed for the nascent Genetic Nurses and Social Workers Union. What better way than to utilise one already in operation? One of the first members of the society was also a member of a sailing club, so they got hold of the sailing club constitution and simply crossed out 'sailing' and substituted 'genetic nursing'.

Counselling was rather more advanced in the USA, consequently when the first master's degree in genetic counselling was set up at Manchester it was by Lauren Kerzan-Storrar who had come across from the USA. By comparison, the story across Europe was highly variable, ranging from well-developed to hardly existing, sometimes genetic counselling having to develop against hostility from medical practitioners who considered themselves the only people qualified to give advice to patients. There is no doubt that it is in countries such as the Netherlands and the UK where health care is very socialised that the integrated service has yielded best results. In the USA there are pockets of excellence, but these seem to be centred around individuals rather than group expertise. An early recognition of a potential problem was that there was an emphasis on a population approach to the control of genetic disease, similar in many ways to the manner in which any genetic research would be carried out. Extending this population approach lead towards a cost versus benefit analysis of testing which takes no account of individual situations and associated stress. As an idea this quite rightly veered away and an individual approach introduced. The former was seen as too near to eugenics where the benefit is to the population not the individual, while the latter was to the benefit of the individual, which in a benign society is what is expected from individuals who take the responsibilities of social membership seriously. This emphasis seemed to develop not from clinical genetics as a whole but with the rise of genetic counselling and its gradually increasing influence on service provision. During these formative years of service development, the Clinical Genetics Society (CGS) produced a number of early documents, such as *Provision of Services for Prenatal Diagnosis of Fetal Abnormality in the UK*; and *The Provision of Regional Genetic Services* in 1978; *Provision of Regional Genetic Services in the UK* in 1982 and in 1983 *Report of the Working Party on the Role and Training of Clinical Geneticists* these were reported in supplements of the *Bulletin of The Eugenics Society*, a situation that many were not entirely happy with, but this was a pragmatic association between two societies. The CGS needed an avenue to publish and the Eugenics Society had both the money and capacity to do so.

Although it was considered to be of immense value to be able to test for genetic disorders prenatally, the idea of amniocentesis, or, indeed, any invasive procedure associated with a foetus, was viewed with great caution. As it was known that many spontaneously aborting foetuses were abnormal, the idea of inadvertently terminating a normal foetus through the agent of an invasive procedure was not something to be considered lightly. The only way that amniocentesis could be developed to the point where it could be offered as a practical option was to test the procedure on patients who were already intending to have an abortion associated with a foetal abnormality. The different methods tried were transcervical and transabdominal. It was the transabdominal route which was the method of choice when it became a more routine technique.

Removal of a small quantity of amniotic fluid for diagnostics was a process which developed very quickly in parallel with human genetics and the human chromosome number. This becomes more apparent when considering that in 1956 Fuchs and Riis working in Copenhagen, Denmark, clearly demonstrated that it was possible to determine the sex of a foetus using cells from amniotic fluid by looking for the inactivated X chromosome, the Barr body, in female cells (Fuchs et al. 1956). The absence of the Barr body is indicative of a male foetus. There was one thing that was needed to reduce the risk to the foetus and mother of the amniocentesis and that was some method of avoiding the foetus and placenta with the needle used to extract fluid. The answer was to visualise the needle as it was being used, which by a process of lateral thinking did become possible.

During the 1930s, high-intensity ultrasound, which was destined to be critical in the development of amniocentesis, was being used in medical practice as a disrupter of tissues and a method of localised heating. It was much later, around 1953, that it became a tool for the treatment of rheumatic arthritis, utilising the controlled heating which could be achieved. In a paper from Gohr and Wedekind in 1940 at the Medical University, Köln suggested that it should be possible to use ultrasound in the diagnosis of abscesses and tumours Gohr and Wedekind (1940). This would depend on the dysfunctional tissues having a different density to the normal surrounding material. It was a little later that Karl Dussik in Vienna first managed to use ultrasound successfully in diagnostics (Dussik 1942). All of this work was associated with soft tissues, but after the Second World War, a joint effort was made between Germany and a team at MIT in the USA to investigate ultrasound as a way of looking inside the skull, hopefully to aid in brain surgery. Unfortunately the project was terminated in 1954. This seems to have been mainly because the equipment was using a frequency of 1 MHz which was really too high, resulting in low penetration of the bones of the skull. It was certainly the case that hard tissue investigations using ultrasound was going to be extremely difficult. On the other hand, however, soft tissues had significant contrast in density, which created a differential echo while at the same time not needing large energies which might in themselves cause damage to the tissue being looked at.

Into the field of ultrasonics came John Wild, a graduate of Cambridge University, who moved to the USA in 1945. He started by looking at ultrasound measurements of bowels after various surgical interventions, trying to follow the healing process. Working for a commercial company, he developed some of the earliest diagnostic equipment, so when he came back to the UK to talk to people, he was listened to as an expert in the field. At one of his lectures, the professor of Medical Physics from the Royal Marsden Hospital, Val Mayneord, was also present with Ian Donald. While he was in the USA, Wild had concentrated on imaging breast tumours, although this was not the route that would prove to be most useful for ultrasound. When Donald left the meeting, he immediately saw the potential of ultrasound in bringing together foetal testing and amniocentesis. The process of changing an idea through experiment and into the clinic was both slow and complicated. In 1958 Donald published a paper (Donald et al. 1958) on the pathologies of abdominal masses as imaged using ultrasound, and the next year the equipment and skill in using it had developed so that he could recognise echoes from a foetal head during pregnancy.

During this time technical development of the equipment, which is an essential part of any modern diagnostic technology, from the microscope to implanted material, was being undertaken in the USA and in the UK by Smiths Industries. After a series of law suits between Smiths and American companies over patent infringements, Smiths withdrew from the market. Donald had been working with them on various aspects of the equipment and so found himself potentially out on a limb. With a great deal of persuasion, he set up his own technical unit at the Queen Mother's Hospital. His newly founded department was very successful in refining the equipment and turning it into a very useful diagnostic tool. In fact the design emanating from this unit became widely used throughout the UK. It was now possible to carry out amniocentesis with only a residual risk to the foetus. There was still a perceived risk, however, simply because of the unstable nature of abnormal pregnancies, which have a very high natural attrition rate, even without any invasive procedures.

When it was felt worth the risk of amniocentesis, the procedure could be carried out successfully, assuming the ultrasound system gave a high enough resolution. These early investigations were most often with the intention of taking a sample, usually foetal blood, for biochemical tests rather than karyotyping. It was in 1966 that Steele and Breg published in *The Lancet* a demonstration that amniotic fluid cells could be cultured in vitro (Steele and Breg 1966). This paper also showed that the cells they extracted originated from the foetus and were therefore suitable for karyotyping. This was confirmed in the same year by a different group. Even so, it was not until 1968 that Nadler published a very significant finding in that he had diagnosed a trisomy 21 foetus from amniotic fluid cells by carrying out a complete karyotype.

It was this move away from diagnosis after the event, be it spontaneous abortion or live birth of an affected child, towards giving the parents a choice by prenatal diagnosis that was the most significant advance in genetic diagnosis. During the early 1960s, a number of discoveries were made, such as the first translocation Down's syndrome in 1960 and the first triploid in 1961, both by Joy Delhanty at the Galton Laboratory in London (Delhanty et al. 1961). These were demonstrating the range of cytogenetic variation, rather than active diagnoses. The discovery of a translocation Down's syndrome is particularly interesting because there was no chromosome banding possible, so with only solid-stained chromosomes, the metaphase can seem annoyingly reluctant to fit together. To the unpractised eye, it can be a very confusing picture which is laid before you as there will be 46 chromosomes, but they will not comfortably sit in pairs. Even stranger would be the status of a male D/G translocation carrier whose karyotype would most easily be seen as a normal female, the translocation product appearing as an X chromosome and the Y being mistaken for chromosome 21 or 22. The realisation that triploidy did occur in humans, if only rarely, was an interesting observation from this period. They do have a distinct phenotype and sometimes go to full term, but survival tends to be measured in hours, presumably due to massive problems of unbalanced physiology. This is hardly surprising as a triploid foetus would have 69 chromosomes, rather than 46.

With the advent of ultrasound and the ability to culture foetal amniocytes, it became possible to introduce carefully constructed screening programmes for those that wanted to take advantage of them. It had been known from very early on that Down's syndrome increased with maternal age, and Lionel Penrose had published this very observation in 1933, although at that time, of course, it was simply not possible to offer any prenatal diagnostic procedures. Since the introduction of prenatal cytogenetic testing, primarily for Down's syndrome, the take-up rate has been considerable. This has had the interesting result that the high-risk group that would have given birth to the largest percentage of Down's syndrome no longer do so. It is the younger mothers who produce most of the live born Down's syndrome children, older mothers having terminations of affected pregnancies. For younger mothers, a further biochemical test based on a maternal blood sample was introduced to assess their risk before amniocentesis was offered. There are two main reasons for this: the first being the inherent risk to a foetus of invasive tests and the second being that the production of a full karyotype from foetal cells is time-consuming, difficult and expensive, so cost for each abnormality found would be disproportionately large. As a by-product of the tests being carried out for Down's syndrome was an occasional surprise, one such was the first antenatal detection of a trisomy 13 in the pregnancy of a 42-year-old.

1967 saw the first practical antenatal study carried out covering the success of various services; there was a reported 90 % cultural success rate in amniotic fluid cell cultures. By the mid-1970s, there had been about 1000 prenatal studies worldwide, which included 50 where there was a recognised D:G translocation in the family.

Alongside the prenatal services provided, it was also seen as desirable to have a blood karyotyping service as well. These started off, as one would expect, under the same roof sharing the same facilities because the skills were essentially identical being based around manipulation of the cell cycle to produce metaphase preparations which could be analysed. Most of the blood samples would be confirmation of chromosomal problems in newborns, after a clinical diagnosis had been made. There would also be a large number of samples looking at parents and families where a bad obstetric history made the possibility of an unusual translocation very real. The development of these services was also seen as sufficiently specialised and of low demand that they could all be housed in one laboratory and cover a large geographical area. In the UK it was normal to have regional laboratories, broadly one to each regional health authority. Although they were usually expected to be under the control of a medically qualified member of staff, some laboratories were more or less autonomous, with the scientific staff not only conducting the tests but collecting samples as well. The one thing that few of these early laboratories had was any sort of formal genetic counselling. Some of the teaching hospitals did have access to genetic nurses, and the medical staff could give some hint to patients regarding simple risk factors in simple cases, but most of the medical staff had only very rudimentary training in genetics; mostly it was knowledge picked up while working and so was skewed in the direction of their consultant interests rather than a global picture of genetics (report of a working party of the Clinical Genetics Committee of the Royal College of Physicians, 1990; Skirton et al. 1997, 1998, 2003).

Early attempts at quantifying recurrence risk seemed to have been stated in such a manner that families often seemed to stop having any more children. This was of such concern that more general phrases about risk and recurrent risk were employed to make understanding easier. With a perennial problem of risk perception, it was never going to be easy to help families with complete and correct data, which might not be available anyway, when they may well already be in a state of stress or shock. Perception of risk rather than objective measure of risk is always problematic, not just in genetics. The same risk can be perceived as either much more likely than it actually is or much less likely, depending upon the mental state of the person involved and their inherent level of pessimism or optimism.

When cytogenetic services started, as E. H. Ford noted, there was a tendency 'to refer all sorts of obscure congenital conditions for chromosome analysis'. Unfortunately this was often pursued as a line of enquiry when it was really of little value, inevitably revealing nothing, while simultaneously putting a great expectation on the outcome for the parents and family. During the 1970s many people imagined that anything unusual could be laid at the feet of genetics and a diagnosis produced. It would also throw an unnecessary burden on the clinical geneticist who would then have an expected negative result and a family with an expectation of precise information. The system rapidly calmed down and stabilised as it became clear that cytogenetics was extremely good at prenatal diagnosis and confirmation of already diagnosed conditions. This relatively small group of conditions included the trisomies, which were broadly recognisable by the clinical staff and only required confirmation. This situation changed with the discovery of more translocation carriers, where an explanation for a poor obstetric history might be forthcoming and sensible advice could be given. Similarly, anomalies involving sex chromosomes were also amenable to explanation and interpretation.

One major area that underwent great development as a result of cytogenetics being available was the study of neoplasias. While it had been recognised for many years that since cells in the body are normally well behaved, for them to go into a spiral of continuous and uncontrolled replication, it would require a change of some sort. The most likely change was obviously genetic, but it was almost impossible to imagine how this worked or how it could be explained. This was especially so with solid tumours where cultured cells often seemed to have random collections of chromosomes, including hyper- and hypodiploidy with what looked like rearrangements amongst the chromosomes as well. It was a reasonable suggestion that some of these were due to cultural artefacts as long-term cultures of these cell lines resulted in degenerative changes in the chromosome complements. Very often these abnormal cell lines would seem to become immortalised, able to break the 'Hayflick limit' of about 40–50 cell divisions before natural senescence took over. In general terms the study of chromosomes in solid tumours was going to be essentially academic until much later on when more sophisticated methods allowed visualisation and understanding of gene rearrangements on specific chromosomes.

The solid tumour investigations were productively confined to the types which were confidently known to show a Mendelian form of inheritance. These constitute about 1–2 % of recognisable neoplasias; numerically that is approximately 50 different types. One of these Mendelian tumours which yielded its chromosomal secret in 1963 was retinoblastoma (Lele et al. 1963). This is a malignant tumour of the epithelial neurological cells of the retina and is exclusively found in children. It is of particular importance because it is not only developmental in occurrence but is particularly treatable by surgery. Lele, Penrose and Stallard described a case with a deleted long arm of a D group chromosome. It was obvious that it was a D group because these chromosomes (13, 14, 15) have a very distinctive shape. As there was no method of banding chromosomes available at the time, it could not be more precisely defined than as being a D group chromosome. Originally it was considered as a coincidental finding, although this was top prove an incorrect assumption. It was later, in 1969, when Wilson described another case in which the long arm of a D chromosome was shown to be deleted in this condition and it was realised that this could be significant (Wilson et al. 1969). A year later it was shown by banding studies to be consistently chromosome 13. There was still a debate as to the association being causal because, as with so many other conditions, the deletion has to be big enough to see, but can be effective when submicroscopic. This is due to it being a single gene defect, so although for reasons not entirely clear deletions are sometimes visible, they are not always so. The need to track these very small deletions encouraged a new range of cytogenetic tests which utilised DNA probes. Indeed, it also powered the search for more accurate delineations of breakpoints in translocations as well as deletions. The manifestation of retinoblastoma is predominantly as a childhood neoplasia, and so just as with Down's referrals, most of the samples originated from paediatric units. The same is also true of many of the leukaemic cases which eventually appeared in cytogenetic laboratories.

Leukaemias have several different aetiologies, but ever since there was a clear definition of a normal histological picture for a blood smear, it has been recognised that when it goes wrong the results are severe.

Although leukaemia could be seen as a hidden disease, cause and effect unknown and result certain, treatment was well known in the nineteenth century. In 1865 Lissauer had already published on the subject, advocating the use of arsenic in its treatment (Lissauer 1865). Although this may seem rather strange, it was seen to be giving relief, but not consistently, and the results were empirical and rather subjective. It fell out of widespread use for many years, only remaining in use in Switzerland where it could still be found as a therapeutic agent into the 1950s. What happened was that the use of such a toxic element was seen as rather basic and so it was superseded by the newer organic compounds. A turning point came in the 1990s when arsenic was found to be used in a modified form of Ayurvedic medicine for the treatment of blood disorders and was investigated as a potential adjunct to other therapies. Another toxic substance which came to be used was urethane. This was not easily taken as it made the patients physically very sick indeed. All of these therapies were piecemeal attempts to control what was to all intents and purposes still a disease without a cause. Probably the beginning of the modern era in chemotherapy started almost accidentally with the realisation after the First World War that it would be necessary to look at the effects of mustard gas so that an effective antidote could be prepared in case it was ever used again. It quickly became apparent that nitrogen mustard produced marked changes in the haemopoietic picture. The peace time possibilities of this were realised, and Cornelius P. Rhoads, head of the medical division of the US Chemical Warfare Service, recruited people that could take on the work once demobilisation had occurred. This is one of the few dual-use cases where the technology was taken from a weapon and moved into therapeutic use. The converse is so often brandished as a reason for not pushing the boundaries of science. Nitrogen mustard, in the form usually referred to as HN2, although very toxic was used as a therapeutic agent in leukaemia and given the name mustine (Fig. 8.1).

One aspect of the disease which did come to mind early on was arrived at by deductive reasoning. Since the major visible aspect of leukaemia was a change in the blood picture away from normal, replacing the blood should help with the associated problems. By giving blood transfusions, it was possible to alleviate the symptoms on a temporary basis. Because this was temporary, it was thought that possibly replacing the bone marrow might be a better option. This was originally tried by Razjock in 1939, but was not really a success except that it paved the way for other attempts. Razjock had given the transplant directly into the bone marrow cavity, but by 1957 a group in Seattle led by Donald Thomas had demonstrated that

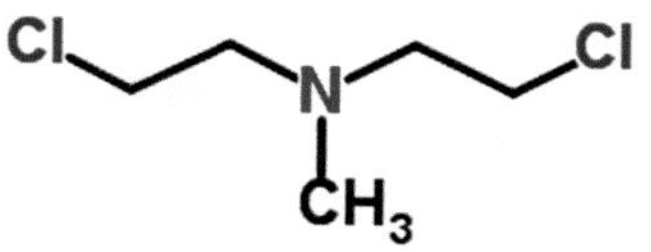

Fig. 8.1 The mustine
molecule, empirically
$C_5H_{11}Cl_2N$. The closely
related mustard gas from
which it was derived has a
sulphur in place of the
nitrogen

large volumes of bone marrow could be infused safely without having to directly access the bone marrow of the recipient (Thomas et al 1957). Throughout the 1960s and 1970s, understanding bone marrow transplants was gradually becoming better. Better still, distinct chromosomal changes were starting to be associated with diagnostically different leukaemia types.

This was set to become much more important in 1960. This was the year in which Nowell and Hungerford found a marker chromosome in their CML cases. Interestingly it was Janet Rowley who determined that it was a translocation between chromosomes 9 and 22 when banding techniques were developed 10 years later (Nowell and Hungerford 1960; Rowley 1973). Before Rowley demonstrated that the small chromosome involved was definitely number 22, it was thought that it might be chromosome 21 which was involved because of the increased risk of leukaemia amongst Down's children. Abnormal chromosomes in leukaemia studies had been reported in 1958, but the quality of preparations and the lack of chromosome banding made accurate determination of changes very difficult. When the report of the chromosome found in CML at the Philadelphia laboratory of Nowell and Hungerford was first published, they intimated that it was considered possible to be something to do with the Y chromosome. This was a reasonable suggestion, but they were slightly misled because all their patients were male. Pat Jacobs had also made slides from CML patients in the UK, but had not started looking at them until the publication of the American results. Her samples also had the marker in them, but some were female, ruling out the involvement of the Y chromosome. It was also Jacobs that dubbed the marker the Philadelphia chromosome or Ph[1] as it was designated. It was given a number because it was expected that more than one such marker would be found. The usual designation now is simply Ph.

Although it took some considerable time for clinical staff to look at these findings with any more than a vague curiosity, they do represent a very specific cause of the disorder, not just a coincidental association. So although leukaemias may be diagnosed and classified by their cytological qualities, it is the recognition of these consistent changes which underlies the disease. By 2001 a new drug, formerly just known as STI571, was approved by the FDA in the USA for use under the name of Gleevec, which preferentially suppresses CML cells with the Ph translocation.

The ability of cytogenetics to characterise the complicated rearrangements found in all manner of leukaemic cell lines also allowed disease tracking to be carried out. In some cases a bone marrow transplant takes place where the recipient has their own bone marrow destroyed and replaced. If this new marrow is taken from a donor of the opposite sex, sex mismatched, it is possible to look for residual disease in the recipient bone marrow by checking the sex of the circulating nucleated blood cells. What is more, this can be carried out on interphase nuclei, so it can be a large and statistically valid investigation for each patient. Although carried out on interphase nuclei, it is still the cytogenetic laboratory where analysis takes place.

References

Andersen DH (1938) Cystic fibrosis of the pancreas and its relation to celiac disease: a clinical and pathologic study. Am J Dis Child 56(2):344–399

Christie DA, Tansey EM (eds) (2003) Genetic testing. Wellcome witnesses to twentieth century medicine, vol. 17. Wellcome Trust Centre for the History of Medicine at UCL, London. Freely available online at www.ucl.ac.uk/histmed/publications/wellcome_witnesses_c20th_med

Clinical Genetics Society, Working Party (1983) Report of the working party on the role and training of clinical geneticists. Bull Eugen Soc 5(Suppl):1–30

Delhanty JA, Ellis JR, Rowley PT (1961) Triploid cells in a human embryo. Lancet 277 (7189):1286

Donald I, Macvicar J, Brown TG (1958) Investigation of abdominal masses by pulsed ultrasound. Lancet 271(7032):1188–1195

Dussik KT (1942) On the possibility of using ultrasound waves as a diagnostic aid. Neurol Psychiatry 174:153–168

Fuchs F, Freiesleben E, Knudsen EE, Riis P (1956) Antenatal detection of hereditary diseases. Hum Hered 6(2):261–263

Gohr H, Wedekind T (1940) Ultrasound in medicine. Klin Woch 19:25

Kazy Z, Rozovsky IS, Bakharev VA (1982) Chorion biopsy in early pregnancy: a method of early prenatal diagnosis for inherited disorders. Prenat Diagn 2(1):39–45

Lele KP, Penrose LS, Stallard HB (1963) Chromosome deletion in a case of retinoblastoma. Ann Hum Genet 27(2):171–174

Lissauer H (1865) Zwei fälle von leucaemie. Berl Klin Wochenschr 40:403

Medical Research Council, Review of Clinical Genetics (1978) A report to the council's cell biology and disorders board by the M.R.C. sub-committee to review clinical genetics. Medical Research Council, London

Medical Research Council, Working Party on Amniocentesis (1978) An assessment of the hazards of amniocentesis. Br J Obstet Gynaecol 85(1):1–41

Miller OJ, Allderdice PW, Miller DA, Breg WR, Migeon BR (1971) Human thymidine kinase gene locus. Assignment to chromosome 17 in a hybrid of man and mouse cells. Science 173:244–245

Nowell PC, Hungerford DA (1960) Chromosome studies on normal and leukemic human leukocytes. J Natl Cancer Inst 25(1):85–109

Old JM, Petrou RW, Modell FK, Weatherall DJ (1982) First-trimester fetal diagnosis for haemoglobinopathies: three cases. Lancet 320(8313):1413–1416

Penrose LS (1933) The relative effects of paternal and maternal age in mongolism. J Genet 27 (1):219–224

Reed SC (1955) Counseling in medical genetics. W B Saunders, Philadelphia, PA

Report of a working party of the Clinical Genetics Committee of the Royal College of Physicians (1990) Teaching genetics to medical students. 24:80–84

Rowley JD (1973) A new consistent chromosomal abnormality in chronic myelogenous leukaemia identified by quinacrine fluorescence and Giemsa staining. Nature 243:290

Simoni G, Brambati B, Danesino C, Rossella F, Terzoli GL, Ferrari M, Fraccaro M (1983) Efficient direct chromosome analyses and enzyme determinations from chorionic villi samples in the first trimester of pregnancy. Hum Genet 63(4):349–357

Skirton H, Barnes C, Curtis G, Walford-Moore J (1997) The role and practice of the genetic nurse. J Med Genet 34:141–147

Skirton H, Barnes C, Guilbert P, Kershaw A, Kerzin-Storrar L, Patch C, Curtis G, Walford-Moore J (1998) Recommendations for education and training of genetic nurses and counsellors in the UK. J Med Genet 35:410–412

Skirton H, Kerzin-Storrar L, Patch C, Barnes C, Guilbert P, Dolling C, Kershaw A, Baines E, Stirling D (2003) Genetic counsellors—a registration system to assure competence in practice in the UK. Community Genet 6:182–183

Steele M, Breg WR (1966) Chromosome analysis of human amniotic-fluid cells. Lancet 287 (7434):383–385

Stevenson AC, Davison BCC, Oakes MW (1970) Genetic Counselling. William Heinemann Medical Books, London

Teaching Hospital of Anshan Iron and Steel Company (1975) Fetal sex prediction of sex chromatin of chorionic villi cells during early pregnancy. Chin Med J (Engl) 1:117–126

The provision of regional genetic services in the United Kingdom (1982) Report of the Clinical Genetics Society Working Party on Regional Genetic Services. Eugenics Society

Thomas ED, Lochte HL Jr, Lu WC, Ferrebee JW (1957) Intravenous infusion of bone marrow in patients receiving radiation and chemotherapy. N Engl J Med 257(11):491–496

Williamson R, Eskdale J, Coleman DV, Niazi M, Loeffler FE, Modell BM (1981) Direct gene analysis of chorionic villi: a possible technique for first-trimester antenatal diagnosis of haemoglobinopathies. Lancet 318(8256):1125–1127

Wilson MG, Melnyk J, Towner JW (1969) Retinoblastoma and deletion D (14) syndrome. J Med Genet 6(3):322

Sex and Chromosomes

While the cogent recognition of sexual differentiation has been around since we were sentient beings, ideas about the way in which sex is determined have been around for a little less time. For most sexually reproducing species, the recognition of the opposite sex requires no intellectual capacity, the dimorphism being quite marked, even to our eyes. Sometimes, however, the recognition of opposites is remarkable. To our eyes many species have no discernible sexual characters, especially amongst birds. For some of these species, for us to determine the sex was more or less impossible without surgical intervention until the discovery of genetic methods.

By the time of Aristotle, there were very many hypotheses as to how sex was determined in humans. The two primary questions which had to be answered regarding sex determination were what mechanism could produce approximately equal numbers of males and females and how the differences are created. With the intrinsic interest in our own sex being paramount, for many years only speculation was made as to the mechanism as experimentation was not going to be possible. Research and guiding ideas would come from work on plants and animals, which would then be extrapolated to the human condition, whether it was correct or not. As ever, the most amenable organisms for research are those with short life cycles, but also there is a need for tissues to be available from which chromosomal preparations could be made. These two requirements rule out prokaryotes, such as bacteria. In most cases suitable cellular material would have to come from tissues so that squashed preparations could be produced; sectioning would not be a reliable method of producing a result, although it was tried in an attempt not just to count the chromosomes but also to determine the relationship of chromosomes to sex determination. Long before chromosomes were looked at with a view to explaining any mechanism involved in sex determination, even before any idea of the cell as a concept, people were keen to explain what seemed both fundamental and an enigma.

When Aristotle put his mind to trying to sort out the conundrum of sex determination, the general assumption was that the sex of a child was determined more or

© Springer International Publishing Switzerland 2016
W.J. Wall, *The Search for Human Chromosomes*,
DOI 10.1007/978-3-319-26336-6_9

less randomly. This was mainly under the influence of Hippocrates, who predated Aristotle by some years, though the dates of his life are rather indeterminate. What Aristotle tried to do was present a specific mechanism that could explain sex determination. Aristotle suggested that the female supplied matter, in itself a nebulous idea, while the male provided form. Joining of the two parts resulted in a point of coagulation which determined the sex of the embryo depending on the superior strength of the male or female part. It may seem as though this was a vague description of events, but it was based on a considerable body of observation and dissections carried out by Aristotle and his students, and it did at least attempt to explain how sexual differentiation comes about. There is no doubt that as an explanation it leaves a great deal to the imagination, but we should not measure the idea against knowledge gained from practical experiments generating data from modern equipment, which would not have been available in any for at the time he was working.

Aristotle set up a school, the Lyceum, at which Theophrastus was a pupil and carried on a considerable amount of scientific investigation. Theophrastus was mostly concerned with plants and significantly recognised the existence of sex in plants; this was way ahead of his time, the idea of sex in plants being conceptually difficult. We think that Theophrastus lived from about 372–287 BC, and yet it was not until the late seventeenth and early eighteenth centuries that a German, Professor Camerarius, described the sexuality of plants and the role of insects in pollination. Even so, the process by which sexual dimorphism was induced remained a closed book. In terms of general ideas regarding sex determination, little movement was made forward in ideas after the classical period. This is not surprising as there was a technological gap which would not be breached for many centuries, and the ideas that emanated from Greece and Rome 2000 years ago held sway with remarkable power.

By the second half of the nineteenth century, the mechanism of sex determination had become a source of speculation on a grand scale, so much that John Arthur Thomson, a naturalist and populariser of science writing at the time, claimed that he had found more than 250 theories regarding sex determination that had all been promulgated in the eighteenth century, all of which were groundless. This was an astute observation because whatever the truth was regarding sex determination, at that time there was nothing substantially supported by experiment and nothing which could be supported by cellular observation.

We now routinely refer to the sex chromosomes in humans as being X and Y, but the way in which they gained this designation is both convoluted and a story taking place over many years. In fact the chromosomes associated with sex determination were both recognised and named long before the total chromosome number was described. The first indirect observation of a sex chromosome which was clearly recorded was by Henking in 1891 using the insect *Pyrrhocoris apterus* (Henking 1891). This is a small black and red bug (Hemiptera:Heteroptera) which occurs widely in mainland Europe. There is an established UK colony, but most records are from migrants arriving in the summer as the British Isles seem to be at the extreme limit of its range. When originally described in the cell nucleus, this observable feature was considered a doubtful candidate to be a chromosome. Even if it was assumed to be a chromosome, it was of unknown action and it was this very uncertainty which was to give it the designation X. Henking was able to show that this strange chromosome only divides during one of the two meiotic divisions, so the primary spermatocyte results in two out of the four sperm containing the X body, while the other two lack it. This system of chromosome division for the X body, by now assumed to be a chromosome, was confirmed in other hemipterans and eventually grasshoppers (Orthoptera). Although in the case of grasshoppers, with such large numbers of chromosomes being involved, it was very difficult to interpret the whole picture. Looking at the chromosomes in females of *Pyrrhocoris*, McClung counted 22 chromosomes in females and 23 in males, so the assumption was made that the accessory chromosome, the X, determined maleness in some way as yet unknown.

It is interesting that it was sex determination where the first suggestion of the relationship between a specific character and a specific chromosome was made. This was a result of observation by McClung published in a seven-page paper in 1901 when the accessory chromosome was said to be male determining. It was in 1905 that a beetle (*Tenebrio*:Tenebrionidae) was used as an experimental organism to demonstrate that there was a small Y chromosome present as well as the X chromosome. This was a clear example of XX female and XY male which was soon demonstrated in many other invertebrate groups (McClung 1901, 1905).

Although by the end of the first decade of the twentieth century it had been demonstrated that there was a direct correlation between what we now know to be the sex chromosomes and the sex of the individual, there was still a question of cause and effect. It was considered possible that the sex chromosomes might not be the cause of the sexual differences which could be seen between individuals; some suggested that they themselves could be a secondary sexual characteristic. It was not possible to rule out at this point that the chromosomal differences were the result of a different and specific sex-determining mechanism, reflected in the chromosome complement. What tipped the balance in favour of accepting that chromosomes were primary causes of sexual differentiation was the straightforward realisation that the presence of an X and Y chromosome system gave an easy and reliable method of arriving at the 1:1 sex ratios that were normally found in animals, most particularly vertebrates. It was considered likely that it was a common feature of sex determination that the heterogametic sex was male and the homogametic sex female until several cases, most notably in Lepidoptera and birds, were shown to be the reverse: heterogametic females and homogametic males. This, then, also gained currency as a possible mechanism of sex determination in man. With this difference of opinion, there were many different attempts to determine the mechanism in man. At this time, material was difficult to come by and preparations poor. Just as the diploid human number was proving elusive due to technological limitations, these also applied to determining anything to do with sex chromosomes. No sensible conclusion could be reached without considerable work and skill. So although a little surprising, regardless of the scientific curiosity, it was human sex determination that remained elusive, while sex chromosome systems were being revealed amongst other animals quite regularly.

Into this cauldron of debate, all manners of potential systems were suggested, based upon the X chromosome being generally recognised as present while all other components were open to debate and interpretation. In 1910 there was a suggestion by Guyer in *The Biological Bulletin* that the human system was XX/O female/male, while 2 years later Gutherz suggested there were no sex chromosomes at all involved in the sex determination of mankind (Guyer 1910; Gutherz 1912).

Although there were many different ideas of varying complexity, one idea which gained considerable purchase on the scientific imagination was the suggestions that sex was determined by the presence or absence of a single X chromosome, no Y chromosome being involved. Without any other background information about how the system functions, this is a valid system because it does allow for a 50:50 sex ratio. It was much later that the XY system was realised to be correct for humans. As a system this would then fit in with most sexually reproducing bisexual organisms and certainly all the mammals. Of course there are exceptions to this, such as the haploid/diploid insects, but these are special cases.

A priest from Silesia, which at the time was part of Prussia, but which is now Poland, called Dzierzon, suggested in 1845 that male honeybees arose from unfertilized eggs and the workers and queens arose from fertilised eggs. This did not receive general acceptance at the time, probably because all the higher animals used in agriculture and kept domestically obviously did not follow this pattern (Dzierzon 1845). Although it did find favour with more experimental work, it was still not a generally accepted concept that sex determination could be haploid/diploid until 1920 when it was shown to be so by Schrader while studying *Aleurodes*, the cabbage white fly (Schrader 1920). Perhaps the most adept cytological investigation on this subject was carried out in 1904 by Meves when he demonstrated that there is a first meiotic division which results in a cell being budded off without any chromosomes, so when meiosis is complete each sperm has an unreduced haploid chromosome complement (Meves 1904). This is a prerequisite of any haploid/ diploid mechanism so that there is always sperm available with a haploid chromosome complement which can be combined with a haploid ovum to always produce a diploid product on fertilisation. All of these various methods which were being discovered and explained differently for sex determination in other species caused some considerable debate, which could be suggested to account for the extensive debate surrounding the balanced sex ratio in humans (Fig. 9.1).

With the understanding that the sex chromosomes in humans came as an asymmetric pair, now routinely designated X and Y, there was a slowing of interest in the more intractable problem of exactly how the presence of these chromosomes influenced the sexual phenotype. This was replaced by increasing interest in determining the total chromosome number, so again this took centre stage. Even so, it was the sex chromosomes which provided a first genetic test based upon a chromosome, rather than a genetic product. The story has an unlikely beginning with two scientists, Barr and Bertram, working at the Medical School of the University of Western Ontario in London, Canada. The original project was

Fig. 9.1 Johan Dzierzon. Known amongst geneticists for recognising parthenogenesis in bees; amongst apiarists Dzierzon is also famous for his extensive research in apiculture

aimed at investigating neuron fatigue in Canadian airmen at a cellular level (Barr and Hamilton 1948). As samples suitable for this were not available from human subjects, the experimental subjects were cats. It was noticed in these preparations that there were nucleolar satellites which could sometimes be seen when neurons were stained using the Nissl technique. This is a method of staining specific areas of cells devised by Franz Nissl, a German neurocytologist. Although he was looking at specific aspects of neuron structure, such as the endoplasmic reticulum and the eponymous Nissl granules, it is a staining technique with wider applications. It uses not just specific stains, such as cresyl violet, sometimes called gentian violet, as well as thionine, but also a specific technique of application to differentially stain the cellular structure under investigation.

What Barr decided to investigate in 1948 was whether heightened activity within neurons caused structural changes which could be seen. After details of the experimental method were worked out, which in itself was quite complicated, Ewart Bertram joined as a graduate student who was studying for an M.Sc. Working with cats, Barr reported that they saw an 'especially prominent mass of chromatin'. Although this was specifically associated with the nucleus of the cells, it was not always present and was only visible in some of the animals. Some of the animals never had this cellular protuberance, which was of interest itself. With the detailed and meticulous records which had been kept, it became apparent that this mass was only clearly seen in females. If it was seen in males, and this was very rare, it was poorly defined. Since staining using the methods they employed can be highly variable, the visibility would also be expected to be variable. Barr was particularly interested in this feature as a possible indicator of physiological activity, although this was shown not to be the case. Barr quickly became aware that this was a sex-related phenomenon, and it was soon demonstrated in the free buccal cells of women but not men. The observation from the feline neurons had been replicated in human epithelial cells (Barr and Bertram 1951).

This observation of the sex chromatin body in female cells would eventually be of considerable consequence in clinical cytogenetics as a method of quickly demonstrating nuclear sex, but would also tie together several lines of thought in genetics which can be clearly seen to start with dosage compensation. This was a phrase first coined by H. J. Muller in 1932 which he used to describe an observation he had made while studying *Drosophila*. This was that in both sexes, regardless of whether there were two X chromosomes (female) or one (male) present, there appeared to be equal expression of the genes which were known to be present on the X chromosome. Of course, so little was known regarding chromosomes in humans that no notion of dosage compensation was possible; in fact it was not always realised that some mechanism for balancing gene product output was going to be required (Muller 1932; Muller and Painter 1932).

There was a tacit belief that since the sex-determining mechanism in humans paralleled that of the fruit fly, *Drosophila*, the mechanism to control X chromosome expression, or more precisely, overexpression, would be more or less the same. So the development of an explanation for the balance of X chromosome expression in humans started with the male/female division of Barr body expression. This was seen in all the cells that were tested with the exception of the germ line cells. For practical purposes, it soon became apparent that root hairs were a good tissue to use. In terms of practical applications of this as a rudimentary method of sex testing, these tests were most frequently carried out on young children, so rather than pulling hair it became normal to use a buccal smear. It was also noticed that polymorphonuclear lymphocytes also showed a sexual dichotomy, with between 2 and 5 % of female polymorphonuclear lymphocyte cells showing what was termed a 'drumstick' attached to the nucleus which was not present in male cells.

Further evidence for the X chromosome being involved in the production of the Barr body came in 1959 when Susumu Ohno observed that one of the two X chromosomes in liver cells of female rats stained more darkly. This is in itself an interesting observation as clinical cytogeneticists who regularly look at hundreds of cells in detail every week will be familiar with this in cultured female cells. Ohno was observing this during late prophase and suggested that since the dark staining (heteropyknotic) chromosome was absent in males, this was the one which could be implicated in the formation of the sex chromatin, Barr, body. In that same year, 1959, it was also demonstrated that mice which had only a single X chromosome, therefore with a karyotype of 39,X, were not only phenotypically female but were fertile as well, the implication being that only a single X was needed for normal development (Ohno et al. 1959).

This picture was demonstrated to be not quite so straightforward in humans when Charles Ford, also in 1959, showed that those individuals described as having Turner's syndrome were phenotypically female, but with only one X chromosome and also lacked a sex chromatin body (Ford et al. 1959). Turner's syndrome had been described in 1938 by Henry Turner an endocrinologist working in the USA. The syndrome was sufficiently well defined phenotypically to be recognisable well before any idea of the chromosomal condition was known. Presentation of these individuals was frequently due to fertility problems (Turner 1938). On the other hand, disrupted sex chromosome distribution in males also showed sex chromatin anomalies. The most widely known sex chromosome abnormality in males is Klinefelter's syndrome which typically have a karyotype of 47,XXY and also have a sex chromatin body, which of course normal males with only a single X chromosome do not. Klinefelter was also an American endocrinologist when he published the paper in 1942 describing the syndrome (Klinefelter et al. 1942).

It was the work of Mary Lyon in England which distilled the observations regarding X chromosomes and their apparent action. Having worked extensively with mice, she was familiar with many of the variant coat colours that appeared from time to time and manifested themselves in different ways depending upon the pedigree. It was obvious that even though not strictly homozygous, males would have a single pattern of colouration, while females might be quite different. The male mice were hemizygous, having only a single X chromosome, and so any coat colour genes that were present would be expressed in all cells. Females, on the other hand, might be heterozygous and therefore be expected to express both genes, which they did not. What Lyon observed was that while some areas manifested themselves as one coat colour form, other areas expressed the other variant. She explained this as being due to one or other of the X chromosomes being deactivated and forming the visible Barr body, a genetically inert X chromosome. Since the colour patterning was in distinct patches, this implied that a further event must be taking place. If X inactivation was occurring in every cell at random, the coat colour would be expected to appear as a blending rather than a spotting of individual colours, as the gene expression changed between random adjacent cells. To explain this, it was suggested that X chromosome inactivation was taking place early on in embryology and every descending cell also inherited the same inactivation pattern of the X chromosome. That way cells might start off with randomly inactivated X chromosomes, but all the daughter cells which made up the surrounding tissue in the adult would have the same inactivation and therefore express the coat colour in a similar way. This original work was published in 1961, but it was a year later that the hypothesis gained traction throughout the Mammalia resulting in the Lyon hypothesis becoming a pivotal idea in dosage compensation (Lyon 1961, 1962).

The major question was to see if this idea could be specifically extended to humans, but here there were less examples of X-linked inherited traits which could be relied upon to demonstrate clonal X inactivation. The problem is a simple one in that while there were already in the early 1960s several X-linked conditions known, they did not generally manifest themselves locally in patches, but acted globally on the entire individual. It was therefore necessary to be a little more circumspect in the interpretation of observation. So disorders such as colour blindness can manifest themselves in reduced sensitivity to red light in females and increased bleeding can be found in women heterozygous for various types of haemophilia. Interestingly, the example of X chromosome inactivation which is most likely to be seen on a day-to-day basis involves cats. Tortoiseshell cats, sometimes called calico cats, are always female as the coat colour is found on the X chromosome, so the inactivation in early embryology of the X chromosomes results in a distinct colour pattern.

By carefully analysing the data, it became apparent that clonal X inactivation was a general feature of mammals and went some considerable way to account for how metabolic balance was maintained in both sexes. It also threw up a further question, which was why, in the murine model, 39X, although short of an X chromosome, resulted in normal fertile adult mice. In humans this was not so. Losing an X chromosome, as in Turner's syndrome, resulted in an infertile adult, as did having an extra one as in the case of Klinefelter's with a karyotype of 47,XXY. This conundrum was answered by Mary Lyon as being due to the time during embryological development at which the switching process occurs.

Once it became possible to determine from cultured cells that a normal male had an XY sex chromosome complement and a normal female XX, it became obvious that things were not always so simple. Turner and Klinefelter had already described their eponymous syndromes, but more variations were appearing than could have been imagined in the early years of clinical cytogenetics. It was a time of persuasive clinical staff and not so many ethical controls, so Klinefelter's patients would often be referred for a testicular biopsy where material could be taken for investigation of sex chromosomes. It was recorded that while testicular biopsies were being taken under general anaesthetic, a sternum bone marrow sample was sometimes taken specifically for somatic chromosome analysis, and by the way it was reported this was not necessarily with the prior approval of the patient. What was discovered was that in broad terms the phenotype depended upon the presence of a Y chromosome, rather than the number or presence or absence of X chromosomes. At the same time, overabundance of X chromosomes, either completely or in mosaic form, rendered the individual sterile. So there were reports of all manner of different conditions, from the classical 47,XXY Klinefelter's to 49,XXXXY and a wide range of combinations in between.

Other multiple X chromosome anomalies seemed to be less disruptive as long as a Y chromosome was not involved. The most surprising was the early discovery of 47,XXX as the women involved were frequently normal intellectually and also in terms of their fertility. When they were first described by Pat Jacobs in 1959, these were dubbed 'super female' in reference to the additional X chromosome (Jacobs et al. 1959). This epithet never gained traction and quickly disappeared from use.

At the opposite end of the scale, there is a syndrome which gave some considerable weight to debates about individuality and the role of genetics in behaviour. The discovery was of an individual with a karyotype of 47,XXY. This was in 1962, but two years earlier, in 1960, a case of 48,XXYY had been described. While 48,XXYY was always going to be extremely rare, 47,XYY was going to prove relatively common. In 1966 Casey and other researchers managed to locate 21 chromatin-positive males by simply looking for sex chromatin (Casey et al. 1966). The study looked at 942 inmates of two hospitals where the majority of patients required additional security due to persistently violent or aggressive behaviour. The two hospitals that were involved in the study were Rampton and Moss Side. These two high-security hospitals are probably not as well known as the rather better known Broadmoor. Rampton is a high-security hospital in Nottinghamshire and was opened in 1912 as an overflow unit for Broadmoor, which is some distance away in Berkshire. The other hospital, Moss Side, was originally a learning disability hospital and was used for victims of shell shock during the First World War. In the 1970s, it combined with Park Lane Hospital to become Ashworth Hospital and became another overflow unit for Broadmoor. The 21 chromatin-positive males were mostly Klinefelter's (47,XXY) syndrome, but a third of them had an extra Y chromosome. Of these seven, five were 48,XXYY and two were 47,XYY/48, XXYY mosaics. This was not a unique finding as in 1965 Pat Jacobs had tested by complete karyotyping 197 out of 303 inmates incarcerated as being mentally subnormal in a similar state hospital at Carstairs in Scotland. Here, nine had aneuploidies associated with the sex chromosomes, seven were 47,XYY, one 48, XXYY and one 47,XXY. The study then moved its attention to a wing of the same hospital which was for psychotic patients. Of the 117 patients out of 139 that were there, two more 47,XXY males were found. Interestingly the feature of these individuals which stood out was their height, all being characteristically tall. The assumption was made that the possession of an extra Y chromosome was associated with tallness and antisocial behaviour. This very simple and unsupported leap was made by the press, because of course it was realised that if general intelligence was affected by Y chromosome, aneuploidy, aggression and antisocial behaviour could be a response to the frustration of unemployment and having to deal with a very complicated society. Also, with respect to this, the natural incidence comes out at about 1/1000, most of whom are perfectly normal members of society.

The sex chromosome aneuploidies can be seen as creating a significant problem, not so much for the individuals affected, or even society, but predominantly for the genetic counsellors who have to explain the many and varied consequences to the patient. A similar dilemma occurs with a particular syndrome with a particular form of expression. This is fragile X syndrome. It was first described in 1943 by Martin and Bell, although at that time they did not consider all of the associated problems which go to make up fragile X as a syndrome (Martin and Bell 1943). What they described was an X-linked mental disability. It was much later, in 1969, that Lubs first reported a marker chromosome which was associated with mental disability. This was only possible when it was a straightforward laboratory procedure to produce metaphase spreads of chromosomes that were good enough to be able to see the fragile site. In this particular instance it turned out to be fragile X chromosomes which were being seen. What Lubs reported in *The American Journal of Human Genetics* was, as he put it, descriptive cytogenetics (Lubs 1969). By investigating a family with three generations of male mental retardation, the cytogenetics could be used to determine that this sex-linked condition was directly related to the strange chromosome which he had found. Initially this was only recorded as a C group, but with the use of tritiated thymidine, it could be shown to be an X chromosome. It was still unknown exactly what was being seen as the fragile site looked very similar to satellites found on other chromosomes. Even though they looked like satellites, it was thought unlikely that they really were satellites in the traditional way because they were large and never shown to be involved in satellite associations, which by then were recognised as normal with D and G group chromosomes. Lubs suggested three different hypotheses for the cytogenetic expression to match the phenotype. The first was that there was a recessive gene closely linked to the secondary constriction, and the second hypothesis suggested that the phenotype was a direct effect of the constriction itself. The third possibility which was put forward was the existence of a variable deletion at the site of the constriction. Because the chromosome preparations were static productions of a dynamic cellular system, it was not obvious that the X chromosome had a fragile site.

It was in the same year Lubs reported his observations of the X chromosome, 1969, that Hecht used the term fragile site for the first time for a chromosome, although in this context he was using it to describe another fragile site, this one being on chromosome 16. Because of the complexity of the system, it would be quite a while longer before a more detailed explanation of what was happening was produced. It turns out that the process is more convoluted than a simple Mendelian mode of inheritance.

Broadly, within this area of the X chromosome, there is a set of base pair repeats, CGG, of which there are normally between 5 and 44, with a modal number of about 30. Sometimes these repeats are tandemly replicated out of line with normal processes, so the system effectively slips and duplication adds CGG repeats. When there are more than 200 repeats, a process of methylation takes place in the repeat of the FMR1 promoter leading to silencing of the FMR1 gene. In this simplified explanation, it becomes obvious why this condition will be fully expressed in males and with variable expression in females as there will be a certain amount of normal X chromosomes which remain fully functional, thereby mitigating the most severe affects.

It is perhaps of some interest that there has been speculation over the years regarding portrayal of fragile X in literature. The reason for this is because fragile X is the most common form of mental retardation at conception after trisomy 21 and certainly the commonest form in males. Following the general form of expression of fragile X, and bearing in mind that details are not generally explored in literature, two characters have been put forward as possibilities. These are Benjy in *The Sound and the Fury* by William Faulkner published in 1929 and Lenny in *Of Mice and Men* by John Steinbeck, published 1937. As can be easily appreciated, this is assuming that these characters were based upon someone the author met or knew, rather than just a general perception of mental disability at the time they were writing.

An interesting aspect of sex chromosome cytogenetics is that the picture they present can give us some insight into the evolution of mammalian groups, specifically the Hominidae. There are 7 extant species of Hominidae in four genera: two species of *Pan*, the chimpanzee and bonobo; two species of *Gorilla*, western and eastern; two species of *Pongo*, orangutans from Borneo or Sumatra; and one species of *Homo*, humans. The picture is complicated by all the members of the family having 48 chromosomes with the exception of man. The complication is that except for chromosome 1, direct comparison with any of our near relatives by chromosome number is misleading. Between the other members of the family, this is not such a problem as their chromosomes all, more or less, match on G-banding. This is because our chromosome number 2 is made up of a fusion of two of the great ape chromosomes. Some of the other chromosomes do have a striking similarity to our own, and an extremely good example of this is the G-banded chromosomes numbers 8 and 9 from *Pongo*. Orangutan has uncannily like chromosomes 11 and 12 from humans. There are many similar examples, the point being that not only are we closely related as species, but the observed structure of the chromosomes which make up our genome is also an observable aspect of our relatedness.

This reflection of relatedness extends to the X chromosome and a lesser extent the Y in a subtle way. The X chromosome banding pattern is almost identical throughout the great apes and so is the pattern of replication. This makes it a very highly conserved chromosome in evolutionary terms.

References

Barr ML, Bertram EG (1951) The behaviour of nuclear structures during depletion and restoration of Nissl material in motor neurons. J Anat 85(Pt 2):171

Barr ML, Hamilton JD (1948) A quantitative study of certain morphological changes in spinal motor neurons during axon reaction. J Comp Neurol 89(2):93–121

Casey MD, Blank CE, Street DRK, Segall LJ, McDougall JH, McGrath PJ, Skinner JL (1966) YY chromosomes and antisocial behaviour. Lancet 288(7468):859–860

Dzierzon J (1845) On the development of bees. Eichstddt Bienenzeitung 1:113

Ford CE, Jones KW, Miller OJ, Mittwoch U, Penrose LS, Ridler M, Shapiro A (1959) The chromosomes in a patient showing both mongolism and the Klinefelter syndrome. Lancet 273(7075):709–710

Gutherz S (1912) Ueber ein bemerkenswertes Strukturelement (Heterochromosom?) in der Spermiogenese des Menschen. Archiv für mikroskopische Anatomie 79(1):A79–A95

Guyer MF (1910) Accessory chromosomes in man. Biol Bull 19(4):219–234

Henking H (1891) Über Spermatogenese und deren Beziehung zur Eientwicklung bei Pyrrhocoris apterus L. Zeit schrift für wissenschaftliche Zoologie 51:685–736

Jacobs P, Baikie AG, Brown WC, Macgregor TN, Maclean N, Harnden DG (1959) Evidence for the existence of the human "super female". Lancet 274(7100):423–425

Klinefelter HF Jr, Reifenstein EC Jr, Albright F Jr (1942) Syndrome characterized by gynecomastia, aspermatogenesis without A-Leydigism, and increased excretion of follicle-stimulating hormone 1. J Clin Endocrinol Metab 2(11):615–627

Lubs HA (1969) A marker X chromosome. Am J Hum Genet 21(3):231

Lyon MF (1961) Gene action in the X-chromosome of the mouse (Mus musculus L.). Nature 190:372–373

Lyon MF (1962) Sex chromatin and gene action in the mammalian X-chromosome. Am J Hum Genet 14(2):135

Martin JP, Bell J (1943) A pedigree of mental defect showing sex-linkage. J Neurol Psychiatry 6 (3–4):154

McClung CE (1901) Notes on the accessory chromosome. Anat Anz 20:220–226

McClung CE (1905) The chromosome complex of orthopteran spermatocytes. Biol Bull 9 (5):304–340

Meves F (1904) Ueber "Richtungskörperbildung" im Hoden von Hymenopteren. Anat Anz 24:29

Muller HJ (1932) Some genetic aspects of sex. Am Nat 66:118–138

Muller HJ, Painter TS (1932) The differentiation of the sex chromosomes of Drosophila into genetically active and inert regions. Mol Gen Genet MGG 62(1):316–365

Ohno S, Kaplan WD, Kinosita R (1959) Formation of the sex chromatin by a single X-chromosome in liver cells of Rattus norvegicus. Exp Cell Res 18(2):415–418

Schrader F (1920) Sex determination in the white? fly (trialeurodes vaporariorum). J Morphol 34
    (2):266–305
Turner HH (1938) A syndrome of infantilism, congenital webbed neck, and cubitus valgus
    1. Endocrinology 23(5):566–574

What We Know, What We Don't and Where This May Lead Us

10

Since the discovery of the human chromosome number in 1956, there have been many developments to take cytogenetics from a research-based science to a highly developed clinical service. Interestingly, one of the areas that might be forgotten in the modern genetic laboratory is looking at chromosomes; the helter-skelter run towards DNA sequencing sometimes misses the target. Cytogenetics remains one of the most technically demanding aspects of modern genetics, mainly because although in the popular mind everything is carried out by computers and computerised equipment, it is still a human activity. We now have digital cameras and sophisticated software which can rearrange chromosomes into a simple ideogram based on size, but the analysis is still carried out by eyeing down a microscope by an individual with 2-year postgraduate training. This is one of the greatest examples of routine pattern recognition that you will come across. Although there is still a great deal to be learnt about chromosome behaviour, it should not be forgotten that it is the contribution that cytogenetics has to make towards understanding the larger subject of genetics which we have to consider.

It is of particular interest that there was a period, one could almost call it a fallow time, when cytogenetics was not regarded very highly. In fact in 1973, E. H. Ford of Cambridge University said that:

> It must be admitted that the cost-effectiveness of the vast sums of money which have gone into human cytogenetics is not at present very great. The contributions to medicine have been considered above, and they do not in aggregate amount to very much, even after 15 years of research. *Human Chromosomes* Page 314. Academic Press, London and New York

Although there is a later acceptance of the importance of cytogenetics to understanding our condition of being human, it was not always so. During this period of the 1970s and 1980s, the attitude voiced by Ford was quite widely held when little could be done with the results of a cytogenetic analysis other than offer a termination of an affected pregnancy or a clearer prognosis in a leukaemia. What

© Springer International Publishing Switzerland 2016
W.J. Wall, *The Search for Human Chromosomes*,
DOI 10.1007/978-3-319-26336-6_10

has happened over a very short period of time is that since 1990 a far more integrated genetics service has made better use of human chromosomes, and human cytogenetics has made use of an increasing body of knowledge of genetics to help develop services. Both now and in the future, it would be expected that cytogenetics would be the starting point for clinical geneticists. This is unlikely to occur as human cytogenetics remains a tricky arena of analysis. In many ways this is unfortunate because analysing high-resolution G-banded metaphases on a regular basis develops a level of insight into the nucleus which may be missed by simple chemical dissection. The trained cytogeneticist is a very clear observer, and while the results can be picked up and the numerical expression understood, arriving at that point may well have taken a team considerable time.

As cytogenetics developed and the resolution of chromosome banding became ever more detailed, the emphasis moved from just looking at chromosomes themselves to trying to explain how they interacted. There is still a considerable literature based around simple observations of clinical cases where a chromosomal rearrangement has been spotted, but these are of limited value as there is little or no attempt to explain the observation. This is much as Lubs described it, a reported natural history rather than anything more insightful. Put another way, there has been a move towards record and report and away from trying to deduce what is really happening within the nucleus as a complete organelle.

This simple reporting of observations may be a reflection of reality or the perception of it, which is gained from wading through uncritical reports of yet another collection of symptoms making up a syndrome. As John Edwards put it in 2001:

> There has almost been no conceptual development in the last 20 years. The development has been technical. The great conceptual development took place in the 1930s, with people like Wright and Hogben and so on, but their papers are all lying about in basements and second-hand bookshops. In—Christie D A, Tansey E M. (eds) (2003) *Genetic Testing*. Wellcome Witnesses to Twentieth Century Medicine, vol. 17. London: Wellcome Trust Centre for the History of Medicine at UCL

In many ways this is true, but while reporting the dull has proceeded, so has the observation of the intriguing and astonishing, born of the same curiosity that drove people to look at chromosomes in the first place. Just considering the size and structure of human chromosomes gives some hint at the astonishing potential of them to behave with great complexity so that they are able to control their own expression through form and function. Do not mistake this as anything other than a product of very complicated interactions between complicated molecules involving many hundreds of feedback control mechanisms interacting in different ways. They, the chromosomes, have no knowledge or sentience.

Estimates of chromosome length and content vary, depending upon who describes them. Part of the problem of making any sense of these statements is that often the estimates of chromosome content are given a huge range. This range simply reflects the very great difference in size between the smallest and largest chromosomes. Trying to measure accurately the content of a structure which is too large and complicated to be analysed using simple biochemical tools and yet too small to be easily handled is never going to be easy. This is why estimates vary by so very much for the DNA content of both diploid nucleus and chromosomes. For example, these are all recorded as for normal human diploid cells.

Author and date	DNA mass	Base pair number	Total length/cell	Length/chromosome
Evans (1982)	12×10^{-12} g	6×10^9	1.9 m	4 cm
McKusick (1969)	5.6×10^{-12} g			
Seuánez (1979)	6×10^{-12} g			
Nature Publications (2001)		3×10^9	2 m	
Rinn and Guttman (2014)			More than 2 m	

It is now generally considered that the mass is around 6×10^{-12} g per diploid cell with about 3,000,000,000 base pairs and a total length of 2 m. The rough guide to DNA length/chromosome would be from 1.4 cm (chromosome 21) to 7.3 cm (chromosome 1).

With the huge variation in heterochromatic DNA present between individuals, these figures will always be estimates. Heterochromatin is generally regarded as being transcriptionally inactive, so content variation in the genome will not affect the cell or individual. The heterochromatic variation seen in some chromosomes can double their size and has been used in family studies before the advent of direct DNA analysis to determine lineage and even paternity in the case of variant Y chromosomes. These cytogenetic studies were quickly superseded by DNA analysis as this is quicker, cheaper and easier.

There is another area of DNA variation which is small and often overlooked but will make it impossible to give precise figures of DNA content for a species, not just *Homo*. This variation is related to ageing. Because of the manner in which DNA replicates, it is necessary to have a chromosomal end cap which does not code for any functional protein. This is the telomere. During the earliest stages of embryology, an enzyme, called telomerase, builds up the length of the telomeres. This is necessary because throughout life, at every round of cell division, DNA replication enzymes will cause a short piece of telomere DNA to be lost. Lose enough and you start losing DNA which codes for essential proteins. By measuring the length of telomeres in cells, it is possible to make a general guess at the age of the cells. The question of course is how telomerase becomes switched on in immortal cell lines such as those found in tumours. If it could be switched off, this could be a useful line of therapeutic attack since the abnormal cells may want to keep dividing but would eventually become compromised by loss of essential genes. Like so many areas of investigation this is rather more complicated than it seems. Control of an enzyme like this is difficult because it is this same enzyme which builds up telomeres in stem cells and keeps them running so that they can, for example, continue to produce white blood cells throughout life by constant cell division without losing their telomeres.

Part of the importance of variation between chromosomes and within genomes was originally termed 'junk' DNA by the biochemists. This was an unfortunate turn of phrase brought about by thinking that there is only function in chemical activity, without considering the possibility that three-dimensional complexity and structural integrity can be just as important.

Simple observation has shown that organisation within the nucleus is an important part of genetics. It was a regular part of the genetics curriculum to make a buccal scrape of cells to look for Barr bodies. The question asked was presence or absence, laudable in itself for clinical purposes, but the bigger question was neglected. Why was it attached to the nuclear membrane so often in the same sort of position? Similarly the appearance of 'drumstick' chromosomes, as originally seen in the neurons of female cats, engendered the question—why should there be a recognisable structure associated with an inactivated chromosome if not because the nucleus is a highly organised structure? Such questions as these probably do not have simple answers. Over the years it has become more than just a suspicion that chromosome structure is important in control of DNA expression. It certainly seems that it is non-coding DNA which seems to be important in holding chromosomes on station so that they can be accurately transcribed. This was given a rather more concrete basis than simple conjecture and deduction in 1989 when it was reported that the nuclear plan was influenced, if not actually controlled, by RNA. The major component which was implicated in this process is nuclear-retained long non-coding RNA (lncRNA). It was the nucleolus which was important in starting this train of thought which led to a practical realisation that chromosomes are an active component in control of genetic expression. A very good example of the potential of this has been found when the information regarding lncRNA is combined with known chromosome activity or, more exactly, inactivity. The Xist gene is known to produce material which silences gene expression, in this case genes on the silenced X chromosome. These are lncRNAs and coat the chromosome which is inactivated; in fact it seems it is the silenced X chromosome which transcribes the lncRNA which controls the silenced X chromosome. This is an example of a very nice simple feedback mechanism.

The history of genetics has been a linear progression from simple questions, why do we look like our parents? Why does a horse give rise to a horse? This also includes simple answers to more complicated questions, such as why do we have chromosomes? For these questions the answer is complicated and they remain mostly unanswered. Early studies, which gave rise to the clear idea that Mendelian inheritance in man was normal, have been superseded, giving rise to a more complicated genetics where the inheritance of a specific gene may be Mendelian, but by the time it has exerted its influence on the phenotype, the picture has become clouded. This should not be such a surprise; we may have a straightforward idea of why a horse produces a horse, but how it is done becomes a question of great complexity.

While we may not be able to restore to balance something as apparently simple as a trisomic condition, this does not mean that it may not be possible in the future. Although this may sound fanciful, it was less than a lifetime ago when the structure of a gene was unknown and now we talk in terms of gene therapy. While it is quite possible that gene therapy may, for all its complexity in theory and practice, be a simple example of the clinical manipulations which will not only be possible but routine in the future, it will be a lot longer before routine treatment for trisomic conditions will be available.

The changes that have gone on within the diagnostics of specified conditions is also astounding, but like many changes to techniques and performance, the changes are not lauded in the same way as breakthroughs, even if they are of greater significance. This is partly due to the public perception of medical tests just being 'done'. That the way a test is carried out and the technology which is used may have changed beyond recognition does not hold any interest for the patient or the general public because the result and the way it is expressed are the same. This is the important factor; how it is arrived at to most people is irrelevant. Only new and better tests are lauded.

Questions in cytogenetics which can be asked with a chance of being answered now include the primary question of why we need chromosomes at all. After all, bacteria do not have them. This is an interesting conundrum where the bacterial observation also has significant repercussions for the interpretation of genetics in humans. While it would be true that the very basic ideas and activities of bacterial DNA are applicable to human genetics, the lack of a constructed chromosome means that the many layers of control which are in place for mammals in general are missing. This has the result that these additional layers of genetic control are sometimes ignored when generalising genetic principles from prokaryotes to eukaryotes.

There must be something about chromosomes which is essential—so much is obvious—or they would not be there. They take energy and resources to construct and maintain, and just as there is a great deal of evolutionary restraint in the structural proteins which control DNA, we can assume that there would be some level of conservation which we could visualise beyond the simple level of biochemistry. For the human condition, this is relatively simple because we have so few close relatives in our family, and in terms of our genus, we are the only extant member. Within our immediate family, there are generally said to be seven species, *Pongo pygmaeus* (Bornean orangutan), *Pongo abelii* (Sumatran orangutan), *Gorilla gorilla* (Western gorilla), *Gorilla beringei* (Eastern gorilla), *Pan paniscus* (bonobo), *Pan troglodytes* (chimpanzee) and *Homo sapiens* (modern man). If the chromosomes of the great apes are looked at using the techniques of G-banding, then we do see considerable similarities. The further removed the relationship is, the less the similarity. This should not be of any surprise as we are looking at a fundamental aspect of being who we are, so as the obvious differences become more manifest, we would reasonably assume that the unseen ones, such as chromosome banding patterns, would as well. Certainly in broad terms, this is the case, with the exception of the X chromosome which is recognised as one of the most highly conserved chromosomes amongst the Mammalia (Charlesworth 1991). So if the general patterns are held stable, but not consistent between organisms, then possibly whatever the mechanisms that require chromosomes to be present would also be consistent. If that is so then it is reasonable to assume that chromosomes would be transferable from one species to another. This is the basis upon which the production of hybrid cell lines is made and why in general they can be stable over many generations with completely alien mixes of chromosomes being present.

It is possible to consider an imbalance, whether a simple aneuploidy or a more complicated mixture of chromosomal material, to be made up of alien material. Strangely and in apparent contradiction of the literal meaning of alien, the additional genetic component may not have originated outside the organism. It is not enough to simply have all the DNA present. It has to be in balance within the cell and in the right order; otherwise the result is not compatible with *Homo sapiens*. So to imagine that the total genome of some artificial cell lines constitutes a human genome would be a mistake. The long established HeLa cell lines are just such an example. These cell lines, of which there are many, were established in 1951 by George Gey at John Hopkins University. They originated from Henrietta Lacks (1920–1951) who presented with an abdominal mass in 1951 at John Hopkins. This turned out to be a malignant cervical cancer which was originally treated with radium inserts. During treatment samples were removed which formed the basis of the HeLa cell lines so extensively used in medical research. Her disease metastasized rapidly and became uncontrollable with the result that the same year she had presented with the condition she died. Henrietta was buried in the vicinity of her mother's grave, where it remained unmarked for many years. Finally, in 2010 a tombstone was erected at her gravesite in Virginia to mark the origin of these extensively used cell lines. One of the first uses of the HeLa cells was by Salk in pursuing a vaccine for polio. This heralded a developing use for these cell lines in biochemical and microbiological research. It was many years before people became curious regarding the chromosomes of HeLa cells, by which time a number of different lines had been established (Skloot and Turpin 2010). Because they were from a malignancy and we now know that degenerate changes take place in such cell lines over time, it would hardly be surprising if these were any different. This is, after all, an immortalised cell line, so there must already have been some genetic changes to the genome to enable continuous uninterrupted cell division. When one of the HeLa cell lines was finally decoded genetically, in this case the Kyoto strain in 2013, an editorial comment in Nature (Callaway 2013) ran:

> The research world's most famous human cell has had its genome decoded, and it's a mess.

What prompted this unequivocal comment was that most of this cell line had one extra chromosome, with some of the cells having five extra chromosomes. Detailed analysis showed that some of the chromosomes had their DNA shuffled and rearranged within the chromosome, implying the DNA was all there, just in the wrong area. Others had their DNA widely distributed around the genome amongst the existing chromosomes. Interestingly, this seemed to have happened quite extensively to chromosome 11, giving rise to the suggestion that since chromosome 11 rearrangements are frequently found in cervical cancers, this may have been an atavistic change associated with the original disease. In broad terms, the cells have become so changed over time that HeLa cell lines are now generally considered to be hypertriploid as they have around about 76–80 chromosomes of various types and configurations present.

The problem for the cytogeneticist when looking at tumour cells is the nature of tumourogenesis. Think of the whole process like a car driving along a cliff top. Everything is perfectly alright until the tyre bursts. This first mutation causes the vehicle to swerve out of control and over the edge of the precipice. On the way down, glass shatters and doors fly off, and the engine cracks and spills its contents. When it comes to rest upside down, it would still be recognised as a car, but no longer of any use as one, so, too, with the cancer cell. Given the time and having been released from constraints of position and differentiation, it develops any number of random problems which makes it impossible to determine from the cell itself what it started out as.

Changes such as those seen in HeLa cell lines are an extreme example of chromosomal changes which can be consistently seen and recorded in tumours. Recording these consistent changes is more important in tumour biopsies than cells from tumours which have been in culture where degenerative changes take place. Long-term tumour growth and metastases in vivo will also cause chromosomal changes which are apparently random as the cells become uncoupled from their restraint on growth and division.

The changes which may be the cause of many tumour types remain enigmatic, but for some the presence of specific rearrangements cannot only be of diagnostic value but aid in treatment as well. One very good example of this is found in some forms of breast cancer. In some cases it has proved possible to use an adjuvant chemotherapy to great effect, but the efficacy is entirely dependent upon the changes involved. An example of this is found where Herceptin (trastuzumab) therapy is seen as appropriate. This monoclonal antibody binds to a specific site on the cell surface called the human epidermal growth factor receptor 2 or when referring to the gene HER2. Perhaps confusingly the gene is also sometimes called Neu or ErbB2, Neu as it is very similar to neu which was cloned from a rodent glioblastoma, a neural tumour and ErbB2 which is similar to avian erythroblastosis oncogene 2. These different gene designations in different species were all shown to be forms of epidermal growth factor (EGFR) genes. Not only that, but later cloning showed that HER2, Neu and ErbB2 were EGFR located on chromosome 17 at 17q12. The significance of this is that normally the measure of Herceptin susceptibility in treatment is based on immunocytochemistry assays carried out on sections of a biopsy which are then scored from 0 to 3 depending upon how dark the staining is, which is directly related to the HER2 receptor numbers on the surface of the cell. This works quite well most of the time, but occasionally the result is equivocal and then the cytogenetic analysis of biopsy samples becomes significant. By using fluorescent in situ hybridisation (FISH), it is possible to count the number of genes and the number of chromosomes of a specific type in a section. The techniques are worlds apart from the nineteenth and early twentieth century explorers of chromosomes, even though, like those early investigators, it still involves looking at chromosomes in paraffin-embedded sections.

FISH works by gently splitting the DNA making up the chromosome using heat and then attaching to the open structure a probe with a fluorescent tag specifically designed to bind to a single gene. Using a probe for HER2, it is then possible to use a microscope to observe probe fluorescing under ultraviolet light and count the number of genes present. Just to make sure the system works, two probes are used, one for the HER2 gene and one for the centromere of chromosome 17. Taking optical sections through the cells, an image can be built up of the number of chromosome 17s present and whether the single gene of interest has been amplified, that is, duplicated along the length of the chromosome to result in overproduction of the HER2 cell surface marker. This is another example where the development of cytogenetics, while crucial for diagnostic purposes, has been very much technology driven. The science is in the probing of genes and the development of clear ideas of how the process of visualisation can be achieved; the technology comes in the digital cameras and very sophisticated microscopes which can operate at different wavelengths and increment the focal plane by tiny fractions to build up a three-dimensional image of a cell. The skill remains with the observer who learns to recognise differences in tissues and count and assess the results generated from the whole process (Fig. 10.1).

Another application of FISH which has made a considerable difference to cytogenetic diagnostics is the use of probes on interphase nuclei for diagnosis of the common trisomies. The difference between these interphase studies and the HER2 studies is that HER2 requires sectioned material, while the prenatal interphase studies use whole cells.

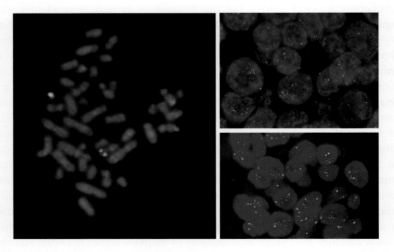

Fig. 10.1 FISH image of interphase nuclei probed for the HER2 gene (*red*) and chromosome 17 centromere (*green*). The multiple copies (amplified) of the HER2 gene can be clearly seen

Previously, it was necessary when looking for trisomies in a foetus to take a sample and culture the cells, either amniotic fluid or chorionic villous. The resultant cells could then be processed and treated to give spread metaphase plates where the chromosomes could be counted and analysed. Although this would occasionally throw up unexpected results, rearrangements and unbalanced translocations, for example, for the majority of cases, it would be a time-consuming process for the laboratory and a highly stressful time for the parents while they waited ten days or thereabouts for a result. Using FISH probes, the situation changed radically. The sample would be collected one day, hybridised overnight with a selection of probes for chromosomes 13, 18, 21, X and Y, and analysed the next morning. The process is quick, relatively cheap and definitely less stressful for the parents. An interesting spin-off of this technique of using interphase nuclei for specific investigations is found in cases of leukaemia in which a patient is given a sex-mismatched bone marrow transplant. If the transplant has been taken and there is no residual disease, then a simple investigation of circulating white blood cells to identify the sex chromosome complement will tell us if there are any of the patient's original cells left. If none of the circulating lymphocytes match the patient's sex, then the transplant is a success.

Without doubt FISH probes have made a considerable difference to the methods that are used in cytogenetic laboratories, but there is always the possibility that we can become enamoured of the technology without making any significant progress in understanding. That would be like using even more powerful computers to play the same games—they may look better but they are still the same. There is even a suggestion that using the complex technology of modern genetics may hinder a deeper understanding of the nucleus and its workings by bringing us back towards a reductionist view of the genome.

In recent years as the specificity of filters and detectors has increased, so has the ability for probes to be discriminated in between, even though their radiant wavelengths may be very close. Similarly, structures of fluorophores have been subtly modified to increase the range of colours available to the investigator. The story does not stop there, because the use of the probes themselves can give increased discrimination. There are three ways in which FISH probes can be used. The first is exemplified by single prenatal diagnosis probes; a single probe is annealed to the target and fluoresced giving a specific wavelength of radiant light. This is the simplest method and the one first used in any practical way in the 1980s. The second method, also from the 1980s, uses a combinatorial system from which a Boolean calculation of the spectral signature of each probe is made based upon the presences or absences of each fluorochrome. This is slightly different to the third technique which is ratio labelling where different probes are labelled with the same fluorochrome combinations, but each probe differs in the proportion of each fluorochrome used as a label. These last two techniques are used to elucidate some of the more complicated rearrangements found in solid tumours and simultaneous visualisation of all the human chromosomes. Multicolour FISH has also been used for screening of small deletions associated with subtelomere sequences.

Since investigating leukaemia is intrinsically easier than solid tumours and since it is possible to produce metaphases from bone marrow samples with relative ease, other methods for solid tumour work have been devised. Probably the best known is CGH, comparative genomic hybridization. With CGH systems it is important to have a normal reference sample of DNA as well as the test sample. Both samples have their DNA extracted and are differentially labelled, say with green or red. By then applying combined probes to target metaphases, amplifications of genomic sequences will fluoresce with the predominant colour of the test sample and deleted sections with the normal probe colour. The ration of colour change can be calculated using the ever more sophisticated image analysis software now available.

CGH is a very useful tool as no prior knowledge is required of the changes that have occurred in the sample, although there are limitations. It does not detect ploidy changes nor does it say anything about structural rearrangements where gains or losses are detected. It is one of the techniques that has led us to realise that many of the changes seen in tumourigenesis are uniform and involve similar if not identical areas of the genome. A development of this technology was described in the 1990s as array CGH. With this technique metaphase chromosomes are not used. Instead the target is made up of large numbers of mapped clones that are attached to a glass microscope slide. Using this technique has increased the resolution for screening genomic copy number changes. The level of resolution is only limited by the clone size and the density which can be put on the slide. By combining G-banding with CGH, it is possible to recognise and be very precise about complicated translocations found in some individuals. The format of array CGH has made it ideal for high-throughput and automated analysis, and like so many of these techniques, it can undoubtedly aid in diagnosis. One drawback of this is that it takes one step away from the cellular basis of mutagenesis and does not help to address some of the fundamental questions about that process. These questions are endless but are primarily associated with the finding of consistent changes in certain tumour types. One such question is why certain changes affect certain cell types, like the Philadelphia chromosome in leukaemia, or why, when the same amplification of a gene is found in different tumours, like HER2 amplification, the response to the same treatment is different. One of the questions associated with the process of tumour development, which could generate considerable knowledge about the workings of our cells and specifically our genome, is exactly why there are hotspots of translocation activity resulting in mutation events within the genome. Of course, it may well be that mutations and rearrangements occur uniformly across the entire genome but only a few result in a viable cell, some of which become cancerous. But if that is not the case, then we must take a step back from a purely molecular investigation of the genome and look again at how it is arranged and constructed within the cell nucleus.

It is possible to use these new techniques in innovative ways to help discern the structure of the nucleus and the chromosomes within it. This is probably going to be of increasing importance as the significance of imprinting and epigenetic phenomena become more obvious and useful in diagnostics. These two closely associated features of cell and tissue development rely on external factors in the control of gene switching and consequent expression. Using many of these new techniques has enabled researchers to separate the idea of high genetic activity from high gene density. As an idea this should be self-evident since there is no intrinsic reason that high density should imply high transcription rates. Alternatively, it has become possible to track changes within the genome in very specific ways. For example, by using a combination of microarrays and chromatin immunoprecipitation, it has been shown that cell senescence is associated with telomere shortening and consequent chromatin damage. These are investigations of the very small scale dealing with specific genetic components in a very large array of genes which remain enigmatic in their control mechanisms.

Using an extension of the techniques developed for such interphase investigations as HER2 diagnostics, it has become possible to fix whole nuclei without disrupting the three-dimensional architecture of the cell. This can then be probed, and a picture in three dimensions built up the position of chromosomes and associated protein structures. As a technique this is another example of technology being essential for the development of new ideas and concepts in genetics and has led to some detailed thinking about the way in which chromosomes are naturally arranged within the nucleus.

It is now obvious, rather than considered probable by observant cytogeneticists who have seen countless metaphases where chromosomes tend to fall together, that chromosomes exist within their own specific domains within the nucleus. These domains are associated with each other, but there seems to be areas where little or no chromosomal activity is found. As an increasing body of knowledge is developing which concerns genetic expression, it seems highly likely that the position within the nuclear domains of whole or parts of chromosomes is pivotal in controlling gene expression. It would also appear that coordination of gene expression between chromosomes when the genes are related in some way is also controlled by a physical positioning of the carrier chromosomes within the nucleus.

This whole and exciting area has been increased and stimulated by the development of FISH probes which can be used on live cells. These are still quite limited in application but have the potential to be very informative to a careful observer. It is becoming increasingly apparent that simply using traditional methods of cytogenetic analysis for anything more complicated than aneuploidy investigations will no longer be acceptable other than on a cost basis. It is also becoming apparent that one of the very oldest questions associated with cytogenetics can now be partially answered. Once the importance of chromosomes was recognised, the question remained, why have chromosomes? The answer seems to have developed, but it is basically that they are the prime vehicles for the control of gene expression and making sure that we all end up looking more or less the same, human.

Cytogenetics teaches us something else about the process of life. Although all eukaryotes have chromosomes and they are obviously essential components in controlling gene expression, we should not assume this to be universal. Should we ever find life on another planet within or outside our solar system, it is extremely unlikely to be compatible with us in biochemical terms. Just because we are all slaves to DNA does not mean that it is the only way that genetic information can be processed and transmitted from generation to generation. Similarly, the histone proteins are highly conserved across species and time on this planet, but that does not mean there are no alternatives. The process of evolution is inconceivably slow, and the development of such complicated organisms, such as ourselves, requires a highly structured nucleus with exquisitely controlled mechanisms of gene expression. But no one sat down and designed it, so to try and imagine an alien alternative to what we have here is a fruitless exercise; what we can say, though, is that it will be different and it will bring with it a whole new era of excitement in genetics.

References

Callaway E (2013) Most popular human cell in science gets sequenced. Nature 495
Charlesworth B (1991) The evolution of sex chromosomes. Science 251:1030–1033
Dennis C, Gallagher R (2001) The human genome. Nature Palgrave, London
Evans HJ (1982) Chromosomal mutations in human populations. Cytogenet Genome Res 33(1–2): 48–56
McKusick VA (1969) Human genetics. Prentice-Hall, Englewood Cliffs, NJ
Rinn J, Guttman M (2014) RNA function. RNA and dynamic nuclear organization. Science (New York, NY) 345(6202):1240–1241
Seuánez HN (1979) The phylogeny of human chromosomes. Springer, Berlin
Skloot R, Turpin B (2010) The immortal life of Henrietta Lacks. Crown Publishers, New York

Appendix A

Rough Guide to Chromosome Structure

Chromosomes are not simple strips of DNA; they are complicated structures based around the central DNA molecule. The thing to remember about chromosomes is that they are not just coding for all the intricate activities of the cell, but they are also coding for themselves. In many ways they are the physical embodiment of the Von Neumann machine. This was a concept originating with John (János) Von Neumann, a perceptive mathematician from Budapest who moved to the USA in 1930. What he proposed was that it should be possible to construct a self-replicating automaton in 1940, although it was not published in its final form until 1966, after his death in 1957. This concept of a self-replicating automaton is perfectly embodied in the physical reality of chromosomes.

To construct a chromosome in its simplest form, there are some components which are required by virtue of either the behaviour of the chromosome or the nature of the chemistry involved. Although we are specifically interested in human chromosomes, this description of chromosome structure and function is true in broad outline for all eukaryote chromosomes. At the chemical level, we have to start with DNA. This is a chain of nucleotides with each nucleotide being a deoxyribose sugar, a phosphate group and a base. It is the base which determines the type of nucleotide. Bases can be either purines (adenine or guanine) or pyrimidines (cytosine or thymidine).

To form the double helix structure, two strands of DNA pair with opposite polarities, the phosphate-sugar backbone on the outside and the bases in the middle being held together by hydrogen bonds. Hydrogen bonds are the only thing holding the two strands together, three between guanine and cytosine and two between thymine and adenine. This bonding formation ensures that pairing always takes place correctly, so if you copy one strand from the other, the original double helix will be reformed accurately.

Along with the DNA, there is a great deal of protein associated with the helix, with an approximate ratio of 2:1 protein/DNA. It should not be imagined that the protein is just floating about; it is associated in a very specific way. There is a mixture of 30 nn histone protein and five histone proteins. Remembering that human chromosomes are very different in size, they also contain different amounts

© Springer International Publishing Switzerland 2016
W.J. Wall, *The Search for Human Chromosomes*,
DOI 10.1007/978-3-319-26336-6

of DNA, from as little as 1.4 to 7.3 cm in length in every nucleated cell. The DNA is packed around a histone protein group, the chemistry of which is quite well known, to make a beaded-string appearance. This is then coiled into a solenoid which is then looped to finally produce the chromosome structure which we know.

So far we have a well-packed continuous piece of DNA in each chromosome, but the final structure must be controlled within the cell, and for this to happen, ancillary structures need to be present. It was during the 1930s that the idea of a chromosome needing an end cap, a telomere, dawned on the scientific community. This was mainly because although X-rays induced mutations in *Drosophila* chromosomes in the form of inversions of the genetic material, if the end of a chromosome was no different to the bulk of the material, it would be expected to see at least some of the rearrangements resulting in the normal ends of the chromosomes being replaced with material normally found further along the chromosome. This was not found and so it was reasonable to suggest that the telomere was in some way protected. It would seem from later studies that the ends of chromosomes do not react in the same way as a broken end would. Observation shows that broken chromosomes have highly reactive ends and tend to fuse with any available free end. Such free end associations can have devastating consequences for an organism producing unstable dicentric chromosomes and even ring chromosomes if both ends of a single chromosome lose its telomeres and stick together.

It transpires that human telomeres, which don't vary much from other species, are made up of lots of repeats based around TTAGGG. What these sequences do is fold back upon themselves and so stop the sticky end problem. Also, because of the nature of DNA replication, they protect the genes within the bulk of the chromosome from being degraded each time the cell replicates.

When a chromosome replicates, there is a little piece of DNA at the telomere which cannot normally be replicated and so is lost. This loss is not normally significant, but does limit the long-term viability of cells to reproduce successfully. It is because of this observation that telomere loss is associated with ageing; although there may not be a causal link between telomere loss and the appearance of ageing, there is certainly a link between age and telomere loss. This, of course, brings us to the inevitable question of why they don't shrink away completely rendering entire species extinct. During meiosis a very specific enzyme appears called telomerase, the sole purpose of which is to add repeat sequences onto the otherwise dwindling telomere. So sperm have far longer telomeric sequences than epithelial cells, because they are rebuilt for the next generation.

The other essential part of a chromosome is the centromere. This is the primary constriction of a chromosome and is characterised by specific repeat sequences. We do not think of humans as being very good experimental organisms, but because of the high throughput in cytogenetics laboratories, apparently rare events can be seen with sufficient frequency for interesting observations to be made. One of the observations from these laboratories is that dicentric X chromosomes, that is, those that have ended up with two centromeres due to unusual end-to-end chromosome fusions, only have one functioning centromere. Such chromosomes demonstrate that the centromere is a dynamic structure, not necessarily a permanent feature. What is essential and associated with the active centromere is the kineto-chore. This is the anchor point for the microtubules at the chromosome end of the spindle involved in migrating the chromosomes into each half of the cell prior to the cell dividing. Without this facility, it would be impossible to guarantee how many of which type of chromosome would end up in each new daughter cell.

Once the components essential to the functioning of a chromosome have been established, there are other features that seem to be more or less essential for the correct transcription and control of the chromosome within the nucleus. These are various types of repeat sequences and interspersed sequences, called introns, that do not directly code for anything, but have a structural purpose. These non-coding sequences seem to be important in anchoring the chromosome in the right place for their function at any given time, attaching them to the nuclear membrane and allowing correct transcription of the DNA into mRNA which then moves out of the nuclear envelope and towards the ribosomes where the data is translated into a polypeptide chain.

Many of these non-coding sequences are associated with disease conditions, such as fragile X syndrome and Huntington's chorea. They are also the system routinely used for paternity testing and in forensic applications of identification.

Appendix B

People in the Text

Abbe, Ernst (1840–1905). Born in Eisenach, Germany, Abbe became professor of physics at Jena in 1870 and director of the astronomical and meteorological observatories in 1878. He was partner with Zeiss and took over responsibility for them in 1888. He produced the Abbe condenser and the achromatic microscope lens in 1886.

Allbutt, (Sir) Thomas Clifford (1836–1925). Born in Dewsbury, Thomas was educated at St Peter's School in York from where he progressed to Caius College, Cambridge. He graduated in natural sciences in 1859, going on to study medicine at St George's Hospital in London and graduating in medicine from Cambridge in 1861. He was elected Fellow of the Royal Society in 1861, became the Regius Professor of Physic at Cambridge in 1892 and was knighted in 1907. Prior to his invention of the clinical thermometer, the thermometers used were about 30 cm in length and took 20 min to register a temperature.

Anaximander (c. 560 BC). He worked at a Greek seaport, Miletus, where he produced a map of the world. He constructed a complicated cosmogony and concomitant zoogony where living creatures emerged from a primeval slime by the heat of the sun.

Aristotle (384–322 BC). He was the son of a doctor in Stagira, northern Greece. He was a member of Plato's academy. He was invited by Philip of Macedonia to educate his son Alexander, retiring to Euboea in 323 BC.

Avery, Oswald Theodore (1877–1955). Born in Halifax, Nova Scotia, Avery studied medicine at Columbia University graduating in 1904 then moving to the Rockefeller Institute from 1913 until retirement in 1948. He did not suggest that genes were simply DNA; that was an idea that emerged after the 1944 publication showing that DNA was the material of heredity.

© Springer International Publishing Switzerland 2016
W.J. Wall, *The Search for Human Chromosomes*,
DOI 10.1007/978-3-319-26336-6

Barr, Murray Llewellyn (1908–1995). Born in Ontario, Canada, Murray originally gained a BA in 1930 and then in 1933 an MD. In 1936 he moved to the University of Western Ontario primarily as researcher in neurology. In 1936 he became a reserve in the Royal Canadian Army Medical Corps and during WWII was in the medical branch of the Royal Canadian Air Force. In 1947 he was joined by a research student, Bertram, and they observed a nuclear satellite that was only present in some of their feline test material. Looking back it was realised these were all females. Eventually it was realised this was the contracted second X chromosome. This is now referred to as the Barr body.

Bateson, William (1861–1926). In 1883 Bateson gained a first-class honours in science from Cambridge, having been described as 'vague and aimless' at school. He worked in the USA and then returned to Cambridge in 1910 where he taught, becoming director of the John Innes Institute which was opened in 1910. He was very interested in evolution and variation and was carrying out breeding experiments when he came across the work of Mendel. He organised the translation and publication in the *Journal of the Royal Horticultural Society*. He coined the word 'genetics', although he never accepted the central nature of chromosomes.

Bell, Alexander Graham (1847–1922). Born in Edinburgh, Alexander was educated at Edinburgh and London and then assisted his father in teaching elocution from 1868 to 1870. He travelled to Canada and in 1871 moved to the USA becoming professor of vocal physiology at Boston in 1873. He specialised in teaching deaf mutes pursuing the idea of visible speech. Most famous for his invention and patenting of the telephone, later developed by Edison, he also started the journal *Science*. Bell was a keen advocate of positive eugenics.

Beneden, Edouard van (1845–1910). Born in Liège, where his father taught zoology, he took over his father in 1870. During the 1880s he demonstrated the constancy of chromosomes in animal cells.

Bertillon, Alphonse (1834–1914). Influenced by his father, Louis-Adolphe Bertillon, who was the head of the Paris bureau of vital statistics, Alphonse developed a complicated system of anthropometry, *bertillonage*, for the identification of individuals. This was completed by 1882 and brought into use by the Paris police. Bertillon became Parisian chief of criminal investigations. Fingerprints were only added later to his complicated system which was used for only a short period of time.

Blakeslee, Albert Francis (1874–1954). Born in Geneseo, New York, Blakeslee was the son of a Methodist minister. He graduated from Wesleyan University with a bachelor's degree in 1896. Following that, he was a teacher at a preparatory school for 3 years before starting postgraduate studies at Harvard University in 1900. He gained an MA in mycology and then in 1904 was awarded a PhD, his work including the important discovery of sexual reproduction in lower fungi. In 1915 he went to Carnegie Institute and by 1936 was director.

Boveri, Theodor Heinrich (1862–1915). Born at Bamberg in Germany, Boveri started his career studying history and philosophy at Munich, but changed to science, graduating in medicine in 1885. In 1893 he started teaching anatomy and zoology at Würzburg. He studied cell division in *Ascaris* and sea urchin eggs. By 1910 his work had made widely acceptable the idea that chromosomes were in some way the vehicles of heredity.

Braconnot, Henri (1780–1855). Born in Commercy, France, Braconnot was apprentice to a pharmacist at the age of 13. At 15 he went to Strasbourg as part of his military service where he worked in the hospital. After a short period working in Paris, he moved to Nancy where he stayed for the remainder of his life. In 1807 he became director of the Botanical Gardens. Most of his work which was carried out in Nancy was in plant chemistry where he discovered the amino acids glycine and leucine.

Bragg, (Sir) William Lawrence (1890–1971). Born in Adelaide, Bragg studied there and at Cambridge originally in mathematics, then moving into physics in 1910. He joined his father (Sir William Henry) in 1912 to research X-ray diffraction. The two Braggs won the 1915 Nobel Prize in Physics, becoming the only father and son to win a joint award and William Lawrence the youngest winner at 25. In 1919 he became professor at Manchester University until 1937 and then in 1938 at Cambridge where he was in charge of the Cavendish Laboratory from 1938 to 1953 where he had Watson and Crick working in the laboratory. He became director of the Royal Institution from 1954 until 1965.

Brown, Robert (1773–1858). The son of an Episcopalian clergyman Brown was born in Montrose, Scotland. He was educated at Aberdeen and Edinburgh. He served in Ireland with a Scottish regiment in 1795 and in 1798 travelled to London where he was appointed to the Flinders expedition of Australia in 1801–1805, bringing back nearly 4000 botanical specimens. He was variously librarian of the Linnaean Society and botanical keeper at the British Museum in 1827, the same year he observed Brownian motion in pollen grains in water, an observation he extended to dye particles. Much later it was recognised as the first direct evidence for the existence of molecules. He observed and named the nucleus in plant cells.

Burrows, Montrose Thomas (1884–1947). Born in Halstead in Kansas, Burrows graduated with an AB from the University of Kansas in 1905 and an MD from Johns Hopkins University School of Medicine in 1905. From there he moved to the Rockefeller Institute before moving to Cornell University in 1911 as instructor in anatomy. By 1915 he had returned to Johns Hopkins where he became associate professor of pathology from 1917 to 1920. From 1920 to 1928 he was a staff member at Barnard Hospital and then Pasadena Hospital until retirement.

Carrel, Alexis (1873–1944). Alexis Carrel was French but spent much time in the Americas before returning to France. He was born in Sainte-Foy-Lès-Lyon and educated at the School of St Joseph and then on to medicine in Lyon. He specialised in vascular surgery but, feeling restricted, was persuaded to travel first to Montreal and then the Rockefeller Institute where he worked on vascular surgery and cell culture, becoming the youngest recipient of the Nobel Prize in Medicine in 1912. During the First World War, he set up and ran a military hospital. With the end of hostilities, he returned to the USA where in conjunction with his friend Charles Lindbergh he invented a device to preserve organs for transplantation. With the defeat of French forces in 1940, he was asked by Marshal Petain to create an 'Institute of Man'. This espoused extreme racism, helping in the deportation of foreign nationals from France. After liberation he was regarded as a collaborator, but died of a heart condition in 1944.

Caspersson, Torbjörn (1910–1997). Born in Motala, Sweden, Caspersson went to the University of Stockholm where he studied medicine and biophysics, one of his interests being the molecular mass of DNA. He received his MD in 1936 and then took a position at the Karolinska Institutet, Stockholm. From 1944 to 1977 he was medical director of the Nobel Institute of Medical Cell research and from 1977 professor in the Faculty of Medicine at the Karolinska Medical Surgical Institute.

Chargaff, Erwin (1905–2002). Born in what was Czechoslovakia, Chargaff studied in Vienna and then Yale, Berlin and Paris and finally in 1935 settled into Columbia University in New York. By 1950 he had demonstrated that while an organism only contained one type of DNA, there were many different types of RNA. Later he demonstrated that the traditional idea of base equivalence was incorrect, while $A = T$ and $C = G$.

Chase, Martha Cowles (1927–2003). Born in Cleveland, Ohio, Martha gained a PhD from the University of Southern California. In 1952, working with Hershey, it was finally demonstrated that DNA was the material of heredity. Her career was made difficult by a developing problem with short-term memory.

Correns, Carl Franz Joseph Erich (1864–1933). Correns was born in Munich and although most well known for being one of the re-discoverers of the work of Mendel, from 1914 he was also the first director of the Kaiser-Wilhelm-Institut für Biologie in Berlin.

Crick, Francis Harry Compton (1916–2004). Crick was educated in North London at Mill Hill and then the University of London where he graduated in physics. His first research was curtailed by WWII where he worked on mines. By 1949 he had moved to the Cambridge Medical Research Unit housed in the Cavendish Laboratory where his skill in X-ray diffraction studies helped in defining the double helix model in 1953, for which he was jointly awarded the Nobel Prize in

1962. He went on to help elucidate the triplet base code. In 1997 he moved to the Salk Institute in San Diego, California.

Darwin, Charles Robert (1809–1882). A slow starter Darwin was unable to make a clear choice of career after 7 years at Shrewsbury School in his home town and 2 years studying medicine in Edinburgh, which he found dull. In consequence his father enrolled him to study for the church at Cambridge. Again the time was not fruitful. When he was 22 he learnt that there was an unpaid position as naturalist on HMS *Beagle*. He wanted to go; his father was against it and his uncle, Josiah Wedgewood, was for it, so in 1831 his adventure started which was to last 5 years. Darwin published his journal of the voyage in 1839. Darwin was living in Kent at Down House with his wife, Emma Wedgwood, who was his first cousin, and their 10 children. 1859 saw the publication of the seminal *The Origin of Species*.

Darwin, (Sir) Horace (1851–1928). Horace Darwin, later Sir Horace 'for services in connection with the war', was the fifth son and ninth child of Charles Darwin and something of a maverick within the Darwin family. He graduated from Cambridge in 1874, returning to Cambridge in 1877 where in 1881 he co-founded Cambridge Scientific Instruments. This was to be a very significant event for biologists. Darwin was essentially the engineer and designer of the company, and so it was no surprise to his friends and colleagues that his office was simply a corner of the work shop. In 1885 Horace designed what was to become a famous scientific instrument, the 'rocker' microtome. This was deceptively simple yet high-quality engineering, allowing serial sections to be cut down to a thickness of only 1/4000th of an inch (0.0006 mm). So robust were these instruments that it was not unusual to find them working reliably for in excess of 50 years.

Davenport, Charles Benedict (1866–1944). Charles Davenport gained his doctorate in 1892. He went on to teach at the University of Chicago and then from 1901 to 1904 was the curator of the zoological museum. Between 1904 and 1934 he was the director of genetics for Cold Spring Harbor Laboratory and founded and ran the Eugenics Record Office between 1910 and 1934.

Democritus (fifth century BC). Native of Abdera, Greece. He elaborated the atomic theory of Leucippus. A lot is said about his personal life, much of which is improbable. Most likely he was from a wealthy family reducing himself to poverty by his extensive travels. It was after he returned home that he gained renown as a thinker and philosopher.

De Vries, Hugo (1848–1935). The son of a Dutch Prime Minister, De Vries was born in Haarlem. He studied at Leiden, Heidelberg and Würzburg. He was the first teacher of plant physiology in the Netherlands and subsequently became professor of botany at Amsterdam from 1878 to 1918. From 1890 he worked almost entirely on matters of heredity, and in 1901–1903 he produced his work *Die*

Mutationstheorie (The Mutation Theory) suggesting that mutations were due to changes in chromosome number rather than alterations to genes.

Dew-Smith, Albert George (1848–1903). Born in Salisbury, Dew-Smith was a wealthy amateur, a member of Trinity College and a student friend of Horace Darwin. Dew-Smith reputedly did not work very hard, but did pass the Natural Science Tripos in 1872. Having been left a considerable legacy in 1870, he was a wealthy man. One outward display of this was that his rooms in college housed a collection of Pre-Raphaelite paintings by such painters as Rossetti, Burne-Jones as well as other of his contemporaries. So well known was he in his social set, both within the university and outside it, that it is said that Robert Louis Stevenson modelled Attwater in his story *The Ebb-Tide* on the real-life person of Dew-Smith. Stevenson probably met Dew-Smith on his many visits to Trinity College where he was an important member of Cambridge society.

Dollond, John (1706–1761). Born in London of Huguenot refugee parents for many years, a silk weaver in 1752, he turned to optics joining his son Peter in his business. John produced a patent in 1758 which paved the way to achromatic telescopes and microscope lenses.

Dollond, Peter (1738–1820). Peter started the optical company which his father, John, joined in 1752. They approached the problem of chromatic aberration, the colour fringes in an image due to differential refraction, which was broadly solved for telescopes.

Donald, Ian (1910–1987). Born in Cornwall, the family moved to Scotland early on. He was educated at Warriston School and the family moved to South Africa where he graduated BA, from the Diocesan College in Cape Town. Moving to London, Donald graduated in medicine from the University of London in 1937. He worked as medical officer with the Royal Air Force Volunteer Reserve in 1942–1946. In 1951 he became reader in obstetrics and gynaecology at St Thomas's Hospital Medical School and then in 1955 professor of midwifery at Glasgow University.

Down, John Langdon (1828–1896). Born in Torpoint, Cornwall, and schooled locally, he was apprentice at age 14 to his father the apothecary. At 18 he went to London where he worked for a surgeon. Down became a student at the Royal London Hospital in 1853 and qualified in 1856. In 1858 he was appointed superintendent of the Earlswood Asylum for Idiots. In 1866 he wrote *Observations on an Ethnic Classification of Idiots* in which he described mongolism, now Down's syndrome.

Dzierzon, Johan (1811–1906). Born in Lowkowitz, in Polish Silesia, as part of a farming family, Dzierzon regarded himself as Polish by birth and upbringing and German by education. He attended a local protestant school before moving to a school in Breslau and then Breslau University, where he graduated in theology in 1833. In 1834 he gained a chaplain's position and a year later in 1835 was ordained as a Roman Catholic priest at Karlsmarkt where he lived for the next 49 years. During his tenure as a priest, he worked as an apiarist and became known as the father of modern apiary. Dzierzon discovered parthenogenesis in bees. It was his work on asexual reproduction in bees as well as his questioning of papal infallibility which resulted in him being excommunicated in 1873.

Edwards, John Hilton (1928–2007). Edwards studied medicine and zoology at Cambridge from 1946 to 1949 and then went on to complete his medical studies at the Middlesex Hospital from 1949 to 1952, when he qualified in medicine from Cambridge. He spent a year as ship's surgeon in the Falkland Islands and then took a post at the University of Birmingham from 1954 to 1956. In 1958 he joined the Medical Research Council Clinical and Population Cytogenetics Unit, returning as professor to the University of Birmingham in 1971, and then in 1979 he was elected Fellow of the Royal Society and became professor of genetics at Oxford.

Fischer, Hermann Emil (1852–1919). Fischer never used his first name and was always known as Emil. Born in Euskirchen, not far from Cologne, Emil wanted to study natural science, but instead went to work for his father, until it became obvious he was unsuitable for the work. In 1871 Emil went to the University of Bonn and then in 1872 the University of Strasbourg, gaining his PhD in 1874. From there in 1875, he went to the University of Munich and then in 1881 was appointed professor of chemistry at the University of Erlangen. From 1885 until 1892 he was professor of chemistry at the University of Würzburg after which he moved to the University of Berlin.

Flemming, Walther (1843–1905). Flemming studied medicine at several different universities throughout Germany finally becoming professor of anatomy at Kiel. He named chromatin and coined the term mitosis. He gave a clear description of chromosome movements but did not know of the work by Mendel and so did not associate the chromosomes with heredity.

Ford, Charles (1912–1999). Originally training in botany, Ford graduated from King's College, London, where he stayed as a demonstrator from 1926 to 1938. Using his knowledge of plant genetics, Charles went to Ceylon, now Sri Lanka, to carry out research on *Hevea*, the rubber tree, from 1938 to 1941 and then 1944–1945, having been interrupted by WWII. From 1946 to 1949, he was at the Department of Atomic Energy Chalk River Laboratories in Ontario, Canada. Upon leaving there, he joined the newly formed Medical Research Council Radiobiology Unit at Harwell in Oxfordshire until 1971. It was here that having started work on plant genetics the copper content of the water from the new pipes stopped the root growth and he switched to animal, and more specifically human, genetics.

Franklin, Rosalind Elsie (1920–1958). Born at Notting Hill, London, and educated at St Paul's Girls' School, Franklin went on to study chemistry at Newnham College, Cambridge, in 1938. After graduating she worked on porosity of coal in the UK and in 1947 went to Paris until 1950 where she learnt the techniques of X-ray diffraction. In 1951 she moved to King's College, London, in the newly formed Medical Research Council Biophysics Unit. It was here that she produced the X-ray diffraction images of DNA crystals. She moved to Birkbeck College, London, to work on viruses until illness stopped her work.

Galton, (Sir) Francis (1822–1911). Born in Birmingham and educated at King Edward's School, Edgbaston, he studied medicine at Birmingham and King's College, London. In 1844 he graduated from Trinity College, Cambridge. He was in every sense a polymath, travelling in North Africa in 1846 and then in 1850 explored unknown territory in South Africa. He published an account of these travels in 1855. Modern weather maps were based on his 1863 publication *Meteorographica*. He also developed fingerprints as method of identification in *Finger Prints* (1892). He researched into colour blindness and developed the idea of the correlation coefficient. He is also well known for his interest in heredity and eugenics which he promoted extensively.

Gamow, George (1904–1968). Gamow was born in Odessa in Russia and was the son of a teacher. He was a student at Leningrad University where between 1931 and 1934 he was professor of physics. He worked in places as diverse as Göttingen, Copenhagen and Cambridge before moving to the USA where he started as professor of physics at the George Washington University and then Colorado University. In 1948 he put forward an explanation for the abundance of chemical elements, most specifically helium. In 1956 he also showed that heavy elements could only originate in the interior of stars. In molecular biology he put forward the idea of a triplet code being able to represent all of the necessary amino acids and stop signals. Although incorrect in details, his ideas were essentially correct as demonstrated in 1960.

Gey, George (1899–1970). Gey was born in Pennsylvania and was the son of German immigrants. He studied at the University of Pittsburgh graduating from there in 1921. He stayed on after graduation and taught zoology at the University. In 1950 he was hired to start the tissue culture laboratory at the Johns Hopkins University where he established the HeLa cells lines which became so important in medical research. He is also credited with the invention and development of the roller drum method of cell culture on solid surfaces. He died of metastatic pancreatic cancer which he had wanted biopsied so that a cell line could be established. His wishes in this were not carried out.

Griffith, Frederick (1877–1941). Born in Hale, Lancashire, UK, Griffith went to Liverpool University after which he worked at the Liverpool Royal Infirmary.

During WWI he became medical officer of the Ministry of Health pathology laboratory, where he became familiar with *pneumococcus* and the different strains.

Haberlandt, Gottlieb (1854–1945). Born in Austria, he was the son of a professor of botany who pioneered soya bean production; his son was also a scientist, the pioneer of hormonal contraception. Gottlieb Haberlandt trained as a botanist, gaining a PhD from the University of Vienna in 1876. In 1877 he moved to Tübingen. By 1880 he had returned to Austria to teach botany in Graz and in 1910 became professor of plant physiology at the University of Berlin.

Hardisty, Roger Michael (1922–1997). Hardisty was born in London and trained at St Thomas' Hospital. It was here that much of his early career took place. He spent a short time working in Cariff and then moved back to London at Great Ormond Street Hospital in 1958. In 1961 he was appointed as head of the Leukaemia Research Unit. In 1969 he became the first professor of paediatric haematology. Hardisty was unusual in giving his home telephone number to worried parents of the children in his care.

Hayflick, Leonard (1928–). Born in Philadelphia, Hayflick went to the University of Pennsylvania, being awarded a PhD in 1956. From there he went to the University of Texas as a postdoctoral researcher. When he returned to Pennsylvania, he spent 10 years at the Wistar Institute and then 2 years at the University as a staff member. In 1982 he went to the University of Florida and in 1988 as professor at the University of California.

Henking, Hermann (1858–1942). Born in Jerxheim, Germany, Henking spent most of his career in applied fisheries research. In 1878 Henking went to Göttingen University to study zoology. After he graduated he stayed on as an assistant in the department until 1892 when he joined the German Fisheries Association. In this position he travelled widely studying fisheries that might be of value to Germany. These trips were as diverse as studying oyster culture in the USA and whaling in Norway. He started an insurance scheme for inshore fisherman and made arrangements for government loans to be made available for the purchase of new motor boats by inshore fishermen. He is best remembered for the discovery of the X chromosome.

Hershey, Alfred Day (1908–1997). Alfred was born in Owasso, Michigan, and went to Michigan State University. He graduated with a BS in 1930 and a PhD in 1934. After Michigan he moved to Washington University School of Medicine until 1950 when he joined the Carnegie Institution at Cold Spring Harbor. In 1969 he was awarded the Nobel Prize in Physiology or Medicine in conjunction with Luria and Delbruck for work on the understanding of the structure of viruses. He became director of the Genetics Research Unit at Cold Spring Harbor, retiring in 1974.

Hippocrates (c. 460–370 BC). Little is known about Hippocrates other than he taught at Cos and travelled widely. Although there are many writing purporting to have come from him, they are unlikely to have done so because over 100 years separate the earliest from the latest. His ideas have become sanctified by time. He broadly described disease as an imbalance of four 'humours' with treatment tending towards rest or exercise and diet.

Hofmeister, Wilhelm Friedrich Benedict (1824–1877). Born in Leipzig he started his working life by following his father into the prosperous music publishing and bookselling business. He developed an expertise in microscopy and interest in botany which led him to demonstrate that the gymnosperms (conifers) are between the cryptograms (ferns) and angiosperms (flowering plants), thereby giving a more unified idea of the plant kingdom. He was professor of Heidelberg in 1863 and appointed professor at Tübingen in 1872. He is regarded as one of the greatest botanists.

Hooke, Robert (1635–1703). Born at Freshwater on the Isle of Wight, Hooke was originally destined for the church but he turned to science. In Oxford he worked for Boyle where he made an improved air pump. In 1660 he moved to London where he was a founder of the Royal Society in 1662. During the 1660s, he devised Hooke's law (as long as the elastic limit is not exceeded, deformation of material is proportional to the force applied). In 1665 he published *Micrographia* describing his compound microscope and used the word 'cell'. It is said that although he was greatly respected, his personality stopped him from being greatly liked.

Hsu, Tao-Chiuh (1917–2003). Hsu was born in Shaoxing in China and was a graduate of Zhejiang University College of Agricultural Sciences. In 1948 he went to the USA where he gained a PhD from the University of Texas at Austin. He remained working in the USA and became president of the American Society for Cell Biology.

Jacobs, Patricia (1934–). Born in London, Jacobs moved to Scotland and studied zoology and botany at St Andrews University. In 1970 she moved to the USA, joining the University of Hawaii in 1972. Returning to the UK in 1988, Jacobs joined the Salisbury cytogenetics laboratory, becoming professor of human genetics at Southampton University.

Janssens, Frans Alfons (1870–1946). Born in Sint-Niklaas, in the East Flanders area of Belgium, Frans was ordained a priest in 1886. He went on to complete a PhD in natural sciences at the Catholic University of Louvain in 1890. By 1891 Frans was teaching at St Lieven Institute, Gent, moving on as professor at the Catholic University of Leuven. He discovered the phenomenon of crossing over in meiosis, which he termed chiasmatypie. This was described in the 1909 publication *La thé orie de la chiasmatypie*.

Klinefelter, Harry Fitch (1912–1920). Klinefelter was an endocrinologist. He started at the University of Virginia and then at Johns Hopkins, graduating in 1937. After staying on for a time after graduation, he moved to the Massachusetts General Hospital, Boston, working there from 1941 to 1942. During the period 1943–1946, Klinefelter was employed in the armed services and then went back to Johns Hopkins University where he stayed until retirement.

Kodani, Masuo (1913–1983). Born in Pasadena, California, he gained a BA in zoology in 1938 from the University of California, Berkley. His wife was a native Japanese, so in 1942 they were forced to relocate to Manzanar, one of the ten relocation centres set up by the USA during WWII. In 1945 he moved to the University of Rochester where he finished his PhD in 1946. From there he moved to the Department of Botany at the University of Wisconsin and at the end of that year moved to Los Angeles and in early 1948 moved to Japan investigating the chromosomal damage found in survivors of the atomic bombs dropped on Japan. In 1954 he moved to the University of Iowa and then the School of Medicine, University of Missouri.

Kossel, Ludwig Karl Leonhard Albrecht (1853–1927). Albrecht received the Nobel Prize in 1910 for his work on nucleic acids. He was born in Rostock where he attended school progressing to the University at Strasbourg in 1827 where he studied medicine. He completed his studies at Rostock, graduating in 1877. Upon graduation he returned to Strasbourg and in 1833 went to Berlin University as director of the Chemistry Division of the physiology department. During the period 1885–1901, he carried out his analysis of nuclein which separated the bases adenine, thymine, guanine, cytosine and uracil.

Lacks, Henrietta (1920–1951). Born in Roanoke, Virginia, in the USA, she was originally called Loretta Pleasant. Upon the death of her mother, she moved in with her grandfather Tommy Lacks. She had her first child aged 14 in 1935 with her first cousin, David Lacks, whom she married in 1941. After giving birth to her fifth child in 1951, excessive bleeding led her to being admitted to Johns Hopkins Hospital where the diagnosis of cervical cancer was made. It was here that a biopsy was taken which resulted in the HeLa cell lines.

Laënnec, Théophile (1781–1826). Born in Quimper, Brittany, he studied medicine at Nantes with his uncle and then later in Paris. He invented the rudimentary stethoscope in 1816.

Lamarck, Jean-Baptiste Pierre Antoine de Monet (1744–1829). The youngest of 11 children, he joined the army at 16. When he left the army on health grounds, he worked in a bank and then studied medicine. In 1781 he became botanist to the King and then professor of zoology in Paris in 1793, after the revolution. Although Lamarckism is poorly considered, he did have an idea of the linearity of descent, if not a correct idea of how characters evolve.

Leeuwenhoek, Antonie van (1632–1723). Born in Delft, he was the fifth child and first son. He never had any formal scientific training and was apprentice to a draper working in Amsterdam until 1650 when he moved to Delft, setting up his own shop and having a paid local government position. He developed his own techniques to polish single lenses of very short focal length. His results were communicated to the Royal Society in 375 illustrated letters. His single lenses magnified from 50x to 200x.

Lejeune, Jérôme (1926–1994). Born in Montrouge, one of three boys, he studied medicine in Paris. He became a researcher at CNRS, Paris, in 1952 and in 1964 the first professor of fundamental genetics at the Faculty of Medicine, Paris. It was at the International Congress of Genetics, in Montreal, in August 1958, that Lejeune discretely mentioned the discovery of a supernumerary chromosome in Down's syndrome. He developed an extreme view of therapeutic abortion in cases of genetic anomalies, being strongly opposed to it.

Leucippus (middle fifth century BC). Probably a native of Abdera, Greece. Although considered the first to suggest an atomic theory, we have only one authentic sentence and no detailed personal information survives. Inextricably linked with Democritus.

Levan, Albert (1905–1998). Born in Göteborg, he studied plant cytology at Lund University gaining a doctorate in 1935. He worked for the sugar industry for a while and became professor of cytology at Lund University in 1961–1973.

Levene, Phoebus Aaron Theodor (1869–1940). Born in Sagor, Russia, he grew up in St Petersburg where he studied medicine. In 1891 he emigrated to the USA and left his family there while he returned to Russia to finish his medical training. In 1892 he went back to the USA to practise medicine. Enrolling at Washington State University in 1894, he investigated the structure of sugars. It was in 1905 after recovering from tuberculosis that Levene was hired as head of biochemistry at the Rockefeller Institute of Medical Research. Besides his scientific endeavours, he was also a considerable linguist being fluent in Russian, English, French and German while also being able to speak Spanish and Italian.

Li, Jonah G (1919–). Li graduated in medicine from the University of California at San Francisco in 1943 and then became assistant in medicine until 1944. He worked at the University of Oregon until 1950 when he became a clinical instructor at the University of California.

Loddiges, Conrad (1738–1826). Born in Hildesheim in Germany, he was a gardener there between 1758 and 1761, at which point the political turmoil of Germany encouraged him to move to England. When he arrived he was taken on as gardener to Sir John Silvester of Hackney. Later he joined another German émigré, John Busch, in a seed company. This was developed until in 1800 and it was the largest plant nursery in Great Britain. In 1818 the nursery introduced the first steam-heated glass house in the UK. The company started importing and selling orchids for which they became famous. The genus *Loddigesia* (Leguminosae) was named in his honour.

Loeb, Leo (1869–1959). Born of German parents in Mayer, Prussia, Loeb was orphaned at a young age and was raised by his uncle. Due to ill health during his childhood, he was educated at a number of schools in spa towns. Eventually he enrolled at Heidelberg University, but spent little time there. He attended for short periods in universities at various German cities. He finally started studying medicine at the University of Zurich Medical School, graduating with an MD in 1897. He carried out some clinical practice at London, Edinburgh and the USA. He moved to the University of Chicago and after a brief period in a clinical position became lecturer at the University of Illinois. In 1910 he moved to the University of Washington and in 1915 became professor there. In 1922, at the age of 53, he married and in 1924 was appointed Chair of Pathology at Washington University, a post he held until his retirement in 1941 at age 72.

Lyon, Mary Frances (1925–2014). Mary was born in Norwich but went to school in Birmingham. In 1943 she went to Girton College, Cambridge, predominantly studying zoology. Having completed her first degree she started a PhD in Cambridge, but then transferred to Edinburgh where the facilities were better. She completed this in 1950 and in 1954 moved to the Medical Research Council Radiobiology Laboratory, Harwell, where much of her ground-breaking work was carried out. Officially retiring in 1990, she remained an active participant in the research at the laboratory until 2012.

MacLeod, Colin Munro (1909–1972). Born in Port Hastings, Nova Scotia, MacLeod started studying at McGill University at 16, graduating in medicine at 23. He was chairman of the Department of Microbiology at New York University where he stayed until 1956. From there he went to the University of Pennsylvania finally returning to New York University as professor of medicine.

Matthaei, Johann Heinrich (1929–). Born in Germany, Matthaei gained his first degree in plant physiology in 1956. He travelled to Cornell University on a 1-year NATO postdoctoral fellowship intending to study cell-free protein synthesis. He worked in the laboratory of Nirenberg, still himself a recent postdoctoral researcher. He returned to Germany in 1962, joining the Max Planck Institute of Experimental Medicine at Göttingen where he is now Professor Emeritus.

Maxwell, James Clerk (1831–1879). The son of a lawyer and schooled at the Edinburgh Academy, Maxwell was nicknamed Dafty. He went to Edinburgh University at the age of 16, having already described a method of drawing ellipses with pins and string. By 1830 he had moved to Cambridge University. After some time as a gentleman farmer in Scotland, he was persuaded to return to Cambridge and set up the Cavendish Laboratory. His many achievements have secured his position as one of the greatest theoretical physicists the world has ever known.

McCarty, Maclyn (1911–2005). Born in South Bend, Indiana, in 1929, he went to Stanford University and then in 1933 medical school at Johns Hopkins University. After graduation he looked for a position and in 1940 he went to New York University and then the Rockefeller Research Institute. He was given his own research laboratory as the Institute became the Rockefeller University.

McClintock, Barbara (1902–1992). Born in Brooklyn, New York, McClintock gained a PhD from Cornell University in 1927, where she developed a special interest in plant genetics, specifically cereal crops. She was sent irradiated cell lines from irradiated maize from Missouri where she saw ring chromosomes. In 1942 she went to Cold Spring Harbor, initially as a temporary member of staff which then became permanent. It was there that she observed transposon activity, the existence of which she was certain of by 1944, although she did not publish her data until the Cold Spring Harbor Symposium in 1951. Later she studied ethnobotany spending many seasons in South America looking at the evolution of corn (maize). She gained a Nobel Prize for her work in 1983.

McClung, Clarence Erwin (1870–1946). Born in Clayton, California, McClung graduated from the University of Kansas in 1892 where he taught from 1893 until 1912 gaining a PhD in 1902. He moved to the University of Pennsylvania as professor of zoology where he stayed until 1940 when he spent a year at the University of Illinois.

Mendel, Gregor Johann (1822–1884). The son of a farmer, he entered the Augustinian monastery at Brno at the age of 21, becoming an ordained priest 4 years later. He studied science in Vienna for 2 years from 1851 after which he started his studies on plant hybridization. This continued until he was elected Abbot in 1868, after which administrative duties did not allow time for botanical research. Although Mendel laid the foundations of modern genetics, he did not succeed in examinations, and it was 16 years after his death before his research made its mark.

Meselson, Matthew Stanley (1930–). Born in Colorado, Meselson was keen on chemistry as a child and set up a laboratory at home. He went to the University of Chicago and then studied physical chemistry at California Institute of Technology in the laboratory of Linus Pauling. By 1957 he finally completed his PhD. In 1954 he was working at Woods Hole and started a collaboration with Franklin Stahl which demonstrated the semi-conservative nature of DNA replication. In the autumn of 1960, Meselson moved to Harvard where he progressed to professor.

Miescher, Johannes Friedrich (1844–1895). Miescher studied medicine in Basel before moving to Gottingen in 1865 to work with Adolf Stecker a chemist. Hi studies were interrupted by typhoid fever which impaired his hearing. He finally graduated in medicine in 1868 but did not feel his hearing made him suitable to pursue a career as physician. His extraction of nucleic acid was so unexpected that his work was repeated before it was accepted for publication. He moved to Leipzig where he became professor. The significance of his extraction of nuclein was not fully recognised until the work of Kossel.

Morgan, Thomas Hunt (1866–1945). Born in Lexington, Kentucky, Morgan studied zoology at Kentucky State College and Johns Hopkins University. He became professor of experimental zoology at Columbia University in 1904, staying there until he moved in 1928 to California Institute of Technology. He remained there until 1945. His most famous works, for which he was awarded the 1933 Nobel Prize in Physiology or Medicine, was carried out using *Drosophila* to develop the theory of specific genes involved in specific tasks being aligned along chromosomes. His later years were spent researching an earlier interest—marine animals.

Muller, Herman Joseph (1890–1967). Born in New York, Muller was very good at school and at 16 entered Columbia University where he gained a BA in 1910. He stayed on at Columbia after graduation for a short time until he went to Cornell to study metabolism in 1911, returning to Columbia in 1912. In 1914 Julian Huxley offered Muller at the newly opened college that was to become Rice University. He moved to Rice for the 1915–1916 academic year, and by 1916 his PhD from Columbia was awarded. In 1918 he returned to Columbia for 2 years until in 1920 he moved to the University of Texas where he stayed until 1932. In 1926 Muller demonstrated a clear quantified correlation between radiation and lethal mutations. 1932 saw Muller move to Berlin and then Leningrad, USSR, in 1933, followed by Moscow in 1934. After a publication which annoyed Stalin, Muller travelled via Madrid and Paris to Edinburgh in 1937 where he remained until 1940 at which point he took up an appointment at Amherst College from where he moved to the Department of Zoology in Indiana University. Herman Muller was awarded the Nobel Prize in 1946 for his work on mutation induction using X-rays.

Nägeli, Carl Wilhelm von (1817–1891). Originally educated in Zürich, he gave up medicine in favour of botany. He studied in Geneva and Jena, eventually becoming professor in Munich in 1857. His views on evolution were broadly Darwinian, although he had a Lamarckian view on the mechanism. He generally considered chromosomes not to be very important.

Nelson, Edward Miles (1851–1938). E. M. Nelson was one of the greatest practical microscopists of the nineteenth century, although his professional career would not lead to that belief. Nelson graduated from Corpus Christi College, Cambridge, after which he joined the telegraph company laying cables along the coast of South America and later the cable joining the Shetland Islands to the mainland. Throughout his career, he was always happy testing the qualities of any optical instruments, such as telescopes and sextants as well as microscopes.

Nirenberg, Marshall Warren (1927–2010). Born in New York, Warren moved with his family to Orlando, Florida, in 1939. He attended the University of Florida at Gainesville gaining a BSc in 1948 and MSc in zoology in 1952. Moving to the University of Michigan at Ann Arbor, he gained a PhD in 1957. From 1957 to 1959, Nirenberg was a postdoctoral researcher at the National Institutes of Health and in 1960 became a research biochemist at the same institute. By 1962 he was head of Biochemical Genetics at the NIH. His interest in DNA started in 1959, with his later collaboration resulting in a Nobel Prize for his work in 1968.

Nissl, Franz (1860–1919). Born in Frankenthal, Germany, his father wanted him to join the church and become a priest, but instead he went to Ludwig Maximilian University to study medicine. In 1884 Nissl began studying brain structure and developed new techniques for staining precise neurological parts. Between 1885 and 1888, he worked at Furstenried castle which had a small research laboratory which he used. Part of his duties were to look after 'mad' Prince Otto. In 1889 Nissl moved to Frankfurt where he tried to relate mental disorders to neurological changes. In 1895 he took up a position at Heidelberg University, becoming professor in 1904.

Nowell, Peter (1928–). Nowell gained a BA from Wesleyan University in 1948 and an MD from the University of Pennsylvania in 1952, where he joined the faculty in the same year. He became chairman of the Department of Pathology at the University of Pennsylvania from 1967 to 1973 and is now the director of the University of Pennsylvania Cancer Centre.

Ohno, Susumu (1928–2000). Born in Seoul, Korea, Susumu was the child of Japanese parents. The family returned to Japan in 1845, where he went to Tokyo University of Agriculture and Technology and gained a PhD in 1949. He moved to the USA in 1952, first as a visiting scholar at the University of California, Los Angeles, and then to the City of Hope Medical Center where he stayed as an active researcher until 1996. In 1970 he published a book, *Evolution by Gene Duplication*, which was highly influential in developing ideas of how evolution could operate at the gene level.

Osgood, Edwin (1899–1969). Osgood joined the University of Oregon Medical School in 1918 gaining an MD in 1924, where he was associated with the biochemistry department and was appointed director of laboratories in 1928. In 1936 he was made the head of Division of Experimental Medicine. His work was primarily associated with abnormal haematology.

Painter, Theophilus Shickel (1889–1969). Born in Salem in Virginia, as a child Painter was regarded as sickly and in consequence was mostly educated at home. In 1904 he entered Roanoke College, graduating in 1908. Painter gained a scholarship to Yale University in 1908 to study chemistry as a graduate student. From there he gained an MA in 1909 and a PhD in 1913. After this he spent a year in Europe before moving back to the USA to work at Woods Hole Laboratory. After working there, he went to the University of Texas in Austin, becoming professor in 1925.

Patau, Klaus (1908–1975). Patau completed a PhD in Berlin in 1936 and then from 1938 to 1939 worked in London and then travelled back to Germany. He worked at the Kaiser Wilhelm Institute for Biology which was situated in Berlin until 1943 when it moved to Tübingen. In 1947 Patau moved to the USA where he became a US citizen in 1948. Working at the University of Wisconsin-Madison, he described the trisomy 13 which became the eponymous syndrome.

Pauling, Linus (1901–1994). Growing up in Portland, Oregon, Pauling was a self-motivated child investigating chemistry at home and school, deciding to pursue it as a career by the age of 15. After attending Oregon Agricultural College, he went to the California Institute of Technology where he completed a PhD using X-ray studies of inorganic crystals, which he completed in 1925. For the next 2 years, he studied in Europe, and when he returned to CIT, he remained in position for the next 35 years. In the 1930s he moved towards organic and biochemical compounds working out structures for proteins. In the 1940s Pauling suggested that the expression of sickle cell disease was due to a change in amino acid sequence in haemoglobin. This was the first time a disease was traced back to its precise molecular origin. He won two Nobel Prizes, for chemistry in 1954 and for peace in 1962, the first person to win two unshared Nobel Prizes.

Perkin, (Sir) William Henry (1838–1907). Perkin started his studies of chemistry in earnest at the age of 15 when he entered the Royal College of Science, despite the opposition of his father. While he was trying to synthesise quinine at home during Easter 1856, he was oxidising aniline. This would not make quinine but did yield mauve. With the aid of his father at the age of 18, Perkin had started a factory making the new dye mauve. After many more successes, chemical and business, he retired at the age of 36 from his dye making to pursue chemical research.

Popper, (Sir) Karl Raimund (1902–1994). Born in Vienna and studying at the university there, he left Vienna in 1937. He moved to Canterbury University College in New Zealand, teaching there until 1945 at which time he moved to the London School of Economics, first as reader and then professor. It was his published works which influenced the ideas of scientific inquiry, one of the most influential being *Die Logik der Forschung* published in 1934. This was translated and appeared in English as *The Logic of Scientific Discovery* in 1959.

Powell, Hugh (1799–1883). Powell started as a supplier to the instrument trade before starting to produce his own microscopes in 1840 which were signed by Hugh Powell. In 1842 he took on his brother-in-law Peter Leland after which his instruments were signed 'Powell and Leland'. On the death of Hugh Powell, his son Thomas ran the company which carried on until the First World War.

Puck, Theodore Thomas (1916–2005). Puck was born in Chicago and except for a short period attended school in Chicago. He also attended university in Chicago, receiving his bachelor's degree, master's degree and PhD from the University of Chicago. His PhD was in physics, and it was the biological implications of ionising radiation which moved him towards genetics and techniques of tissue culture.

Quekett, John Thomas (1815–1861). John came to London as apprentice to his brother Edwin. He went on to the London Hospital Medical College and King's College, qualifying in 1840. Once qualified, John went to the Royal College of Surgeons. In 1848 he wrote *A Practical Treatise on the Use of the Microscope* which was revised in 1852 and 1853. In 1857 he was elected Fellow of the Linnaean Society and in 1860 the Royal Society. When he was approached to become president of the Royal Society, he wrote to postpone the decision due to illness, but unfortunately the letter was delayed and upon his recovery he discovered he had been elected president in his absence. With his brothers William and Edwin, the Queketts started the Quekett Club for microscopists.

Robiquet, Pierre Jean (1780–1840). Born at Rennes in France, Robiquet was schooled at Château-Gontier, the school being closed before he finished his education due to a dispute between the religious teaching staff and the civil authorities. He became apprentice to a pharmacist in the town of Clary. During the French Revolution, he was an army pharmacist becoming professor at the École de Pharmacie. Besides his work on amino acids, in 1832 he was the first to extract codeine.

Roux, Wilhelm (1850–1924). Born in Jena, Roux was the fourth son of a well-known fencing master. His education at Jena University was interrupted in its first year by the Franco-Prussian War. Upon his return to education, he matriculated from the Department of Medicine in 1873. He completed his dissertation in 1877 and passed his state medical examinations in 1878. His first employment was in Leipzig in 1879 after which he moved to the Anatomical Institute at Breslau until 1889 when he moved to Innsbruck becoming professor of anatomy. In 1895 Roux became the director of the Anatomical Institute of the University of Halle, a position he held until 1921.

Salk, Jonas (1914–1995). Born in New York, Salk was one of three brothers, children of Polish immigrants. Recognised as a gifted scholar, he was educated locally and enrolled at the City College of New York at age 15 where he graduated with a bachelor's degree in chemistry in 1934. After this he went to New York University School of Medicine. After graduating he went to work at Mount Sinai Hospital and in 1941 spent 2 months in the virology department at the University of Michigan; this set him on a path towards virology research. In 1947 he joined the University of Pittsburgh School of Medicine which is where he developed the eponymous polio vaccine.

Schleiden, Jakob Matthias (1804–1881). Born in Hamburg, Schleiden studied law at Heidelberg. He practised law for a while, but his interest in botany became overwhelming and eventually he graduated from Jena in 1831. By using the modern microscopes then becoming available, he studied plant structures and finally convinced that all plant structures were made up of cells. With Schwann he developed the cell theory. He was a popular lecturer and debater on scientific matters.

Schott, Otto (1851–1935). Otto Schott invented borosilicate glass. He started his career studying chemical technology in Aachen and then Wurzburg, Leipzig and Jena. In 1879 he developed his borosilicate glass. This led him in 1884 to join with Zeiss and Abbe to set up Schott and associate Glass Technology Laboratory in Jena.

Schwann, Theodor (1810–1882). Schooled in Köln, Schwann studied medicine in Berlin and graduated in 1834 where he stayed on as assistant to Müller. He discovered the Schwann cell, a myelin sheath around peripheral nerves, and that an egg is a single cell. He is best known for the cell theory. Schleiden had argued that plants were all based on cells, and Schwann developed this for animal material as well. This was defined in a book written by Schwann in 1839, a year after he had left Germany for Belgium thinking he had no career prospects in Germany. In Belgium he became a recluse and mystic and more or less stopped doing science.

Seysenegg, Erich Tschermak von (1871–1962). Seysenegg was an Austrian agronomist whose main work was breeding disease-resistant crops. He gained a doctorate from Halle in 1896 and was teaching in Vienna in 1901. He was one of the re-discoverers of Mendel's paper in 1900. It is interesting to note that his maternal grandfather was an academic botanist who taught Mendel in Vienna.

Stahl, Franklin William (1929–). Brought up in Needham, a suburb of Boston, he went to Harvard College and then Rochester University where he completed his PhD in 1956 studying T4 phage. From 1955 to 1958, Meselson carried out research at California Institute of Technology and from 1958–1959 at the University of Missouri. In 1959 he moved to the Institute of Molecular Biology at the University of Oregon, Eugene. In 1964 he demonstrated that T4 phage has circular DNA. He retired in 2001.

Strasburger, Eduard Adolf (1844–1912). Strasburger was born in Warsaw and studied botany in Paris, Jena and Bonn. He spent from 1869 to 1880 teaching at Jena, and then from 1880 to 1912 at Bonn, he was an enthusiastic Darwinian. His description of mitosis in his book *Cell Formation and Cell Division* (1875) was clear enough for him to conclude that the nucleus was the centre of heredity. He demonstrated that the passage of sap in plants was due to capillarity rather than an active physiological process. His textbook, written in conjunction with others, *Strasburger's Textbook of Botany*, was widely used through the twentieth century.

Sutton, Walter Stanborough (1877–1916). Born in Utica, New York, Sutton was the fifth of seven sons. He was raised on a farm in Russell, Kansas, where he went to school, after which in 1896 he attended the University of Kansas originally to study engineering. He switched to biology in 1897 and graduated in 1901. He moved to Columbia University to study zoology but did not complete his PhD, returning to work on the Kansas oil fields for 2 years at age 26. He returned to Columbia University in 1905 where he wrote *The Chromosomes in Heredity* before moving to the University of Kansas as professor. He died aged 39 from complications of appendicitis.

Svedberg, Theodor (1884–1971). Born in Valbo, Gävleborg, in Sweden, Svedberg joined Uppsala University in 1904, where he remained for his entire career. It was his intention to apply chemical methods to analysis of biological problems. One of his projects which became of inestimable importance was the ultracentrifuge. He was awarded a Nobel Prize in Chemistry in 1926. The unit of sedimentation velocity is named after him, the Svedberg (S).

Talbot, William Henry Fox (1800–1877). Talbot was educated at Harrow and Trinity College, Cambridge. In 1839 he described 'photogenic drawing', or photography, prints on silver chloride paper. In 1841 he patented the first process of making a negative from which a print could be made, described as a calotype. In 1844 he produced the first book illustrated with photographs, called *The Pencil of Nature*. He also helped in deciphering cuneiform inscriptions from Nineveh.

Thomson, (Sir) John Arthur (1861–1933). Born in East Lothian in Scotland, Thomson became a great populariser of science as well as a serious scientist. He taught at the Royal Veterinary College from 1893 to 1899 when he went to the University of Aberdeen as professor of natural history where he specialised in the study of soft corals. He was knighted in 1930.

Tjio, Joe Hin (1919–2001). Tjio was born in Pekalongan, Java. He was educated in Dutch colonial schools and trained in agronomy. During the Second World War, he was intern in a Japanese concentration camp. After the end of the war, he went to the Netherlands and over the next few years he worked n Denmark, Spain and Sweden. From 1948 to 1959 he worked in Zaragoza. He moved to the USA and joined the National Institutes of Health at Bethesda in 1959. He gained a PhD from Colorado University and retired in 1997.

Turner, Henry Hubert (1892–1970). Turner was an American endocrinologist who published the description of the eponymous syndrome in 1938. Born in Harrisburg, Illinois, he graduated in medicine from the University of Louisville School of Medicine in 1921. After a brief period of postgraduate training in Vienna and London, he returned to the USA and joined the University of Oklahoma College of Medicine where he spent most of his career.

Turpin, Raymond (1895–1998). Turpin had just completed his medical studies when he was mobilised in 1915 and was sent to the fortresses of Verdun. After the First World War, he turned to paediatrics and was involved in the first trials of vaccination against childhood TB. In 1930 he turned to hereditary diseases and in 1941 introduced teaching genetics into the Faculty of Medicine at Paris University. In 1958 he created the first chair of fundamental genetics in France.

Vauquelin, Louis Nicolas (1763–1829). Born in St André d'Hébertot in France, Louis worked in the fields alongside his father doing well at school and becoming apprentice to an apothecary at the age of 14. He worked as an apothecary in Rouen and Paris, where he was taken on by a chemist, de Fourcroy. Vauquelin had to leave Paris temporarily in 1793 when he rescued a Swiss soldier from a mob during the French revolution. He is best known for his discovery of chromium and beryllium, but in biology it was his reputation as the first person to isolate an amino acid, asparagines, which he extracted from asparagus.

Waldeyer-Hartz, Wilhelm (1839–1921). After studying science Waldeyer-Hartz (often shortened to Waldeyer) graduated in medicine and moved to Berlin where he taught anatomy and physiology. He introduced the words 'chromosome' and 'neuron'. His account of the spread and development of cancer in 1863 reads as surprisingly modern, concluding that cancer starts as a single cell, spreading by migration through the blood or lymphatic system.

Wallace, Alfred Russel (1823–1913). Wallace left school at 14 and started work as a surveyor and then became a teacher in Leicester. His first expedition, with H. W. Bates, to South America ended poorly with his ship with his samples onboard being destroyed by fire. In 1854 he embarked on an expedition to Malaya where he developed ideas very similar to those of Charles Darwin. Wallace became a leading advocate of Darwin's ideas. He created the ideas of zoogeography recognising the importance of geology and geography in separating flora and fauna. Wallace's line is an imaginary line running between the islands of the Malay Archipelago separating oriental fauna from Australian fauna.

Watson, James Dewey (1928–). Born in Chicago, Illinois, he attended Horace Mann Grammar School and South Shore High School. This was followed by the University of Chicago where he enrolled at age 15. He gained his BS zoology in 1947 and a PhD from Indiana University in 1950. In September 1950, Watson spent a year at Copenhagen University, after which he moved to the Cavendish Laboratory, Cambridge, and carried out the research for which he received the Nobel Prize in 1962 in conjunction with Crick and Wilkins. In 1956 Watson moved to Harvard University where he eventually became professor. In 1968 he became director of the Cold Spring Harbor Laboratory. In 2007 Watson was suspended following criticism of views on race and intelligence attributed to him, later retiring aged 79.

Whitworth, (Sir) Joseph (1803–1887). Whitworth was an engineer born in Stockport, Cheshire. His engineering skill was extensive, producing the standard thread for attaching objective lenses to microscope tubes. In 1859 he invented a gun of compressed steel with a spiral polygonal bore. For many years Whitworth's eponymous thread was a standard in engineering.

Wilkins, Maurice Hugh Frederick (1916–2004). Wilkins was born in New Zealand, but by the time he was 6 years old, the family moved to Birmingham. He was educated at King Edward's School from 1929–1935, after which he attended St John's College, Cambridge, where he gained a BA in physics. After that his continued education led to a PhD from the University of Birmingham. During the Second World War, Wilkins worked at Birmingham on radar screens and on the Manhattan Project at the University of California, Berkeley. In 1945 he moved to St Andrews University Department of Physics. When the head of department moved to King's College, London moved to set up a new biophysics department and Wilkins also moved to become assistant director. During the 1930s he was a member of Cambridge Scientists Anti-War Group.

Winiwarter, Hans von (1875–1949). Hans was born in Vienna and was 3 years old when the family moved to Liege in Belgium. He received an artistic and musical education at home, developing a taste for Japanese Ukiyo-e engravings and amassing a considerable collection. He was the eldest son of the family and became a doctor of anatomical science in 1910.

Wollaston, William Hyde (1766–1828). Wollaston came from a family of scientists and physicians, a route he followed, studying in Cambridge and London. In 1800 he gave up his medical practice and took up a partnership with Tennant to produce and sell platinum. It was during this period that he discovered the elements palladium (1802) and then rhodium (1804). Although he published his discovery of rhodium in the usual way, palladium came to the attention of the public simply by being offered for sale. Wollaston made a considerable fortune from the sale of malleable platinum, which had not previously been available. He kept the secret of its manufacture until near his death. He also worked extensively in optics, designing many improved measuring devices.

Woodhull, Victoria (1838–1927). Born as Victoria Claflin in Homer, Ohio, she was one of ten children of a family making a living as fortune tellers. She specifically worked with her younger sister, Tennessee. She married Dr. Canning Woodhull in 1853, but divorced in 1864. In 1868 she went with Tennessee to New York and persuaded Cornelius Vanderbilt to set them up as stockbrokers. At this time he started advocating free love, equal rights and legalised prostitution. Victoria won support via the women's suffrage movement to become the first women nominated for presidency of the USA. In 1877 she moved to London where she wrote *Stirpiculture, or the Scientific Propagation of the Human Race*, published 1888.

Zeiss, Carl (1816–1888). Carl Zeiss was born in Weimar on 11 September 1816, the son of a toy shop proprietor. He became apprentice to Friedrich Korner, *Hofmechanikus*, literally translated as official mechanic; at the University of Thumingen in Jena, Carl Zeiss learnt many skills and went on to work with instrument makers in Stuttgart and Vienna. When he first started his workshop, he was on his own, not only making instruments but also repairing them. By 1847 he had introduced a single-lens microscope for sale. Being a single lens probably demonstrated his awareness of magnified aberrations in compound instruments as well as the essential portability of single-lens devices. They would, of course, also be both easier and cheaper to manufacture. Business improved with time, and he moved premises in 1858 and again 20 years later to even bigger ones. By this time he was recognised as making the best lenses in Germany.

Zernike, Frits (1888–1966). Born in Amsterdam, both parents were teachers of mathematics. He was regarded as very good at science at school. He joined the University of Amsterdam in 1905 where he studied chemistry. He was awarded a prize for his work on opalescence in gases in 1912, which went on to form part of his PhD thesis in 1915. In 1913 Zernike was assistant at the astronomy laboratory of Groningen University, where in 1915 he was appointed to a position in the physics department and in 1920 as professor of theoretical physics. In 1930 he started investigating the spectral line anomalies associated with diffraction gratings and by 1933 described the phase contrast techniques at a physical and medical congress in Wageningen. He was awarded the Nobel Prize in Physics in 1953.

Glossary

Agglutination If antibodies attach to antigenic sites on cells, the adjacent, divalent antibodies will form bridges and attach the cells together. This is the basis of agglutination tests for blood groups.

Allele Variant of DNA sequence in a single gene. In diploid cells there will be two alleles for each gene and in haploid cells a single allele for each gene.

Amnion Foetal membrane which forms a sack containing the amniotic fluid in which the foetus is suspended.

Amniocentesis Process of removing a small quantity of amniotic fluid from around the foetus. This is a technique usually carried out using ultrasound to guide the positioning of the extracting needle.

Amniotic Fluid Fluid surrounding the foetus and of foetal origin. This serves as a support in utero and contains a changing composition of organic molecules and cells, depending upon the time of gestation. The cells which can be found in the amniotic fluid can be extracted by amniocentesis to be tested for a variety of genetic and biochemical abnormalities and can be grown to produce a karyotype of the foetus.

Aneuploid A somatic cell where the chromosome number is not completely diploid, there being an extra chromosome present, or missing. Down's syndrome is an example with the extra chromosome 21 and Turner's syndrome, 45X, with one X missing.

Antibody Protein molecules only found in vertebrates and are synthesised by B cells and able to attach to specific antigens. Antibodies are a type of immunoglobulin.

Antigen Substance that cause an immune reaction.

Apomixis Asexual reproduction in plants without pollination. All plants arising from apomixis constitute a clone. This is quite different to self-pollination.

Autosome Any chromosome which is not specifically associated with sex determination. Human chromosomes 1–23 are autosomes. X and Y are the sex chromosomes.

Ayurveda A medical system, part of the tradition of holistic treatment, based on the idea that health is based on a balance between mind, body and spirit.

© Springer International Publishing Switzerland 2016
W.J. Wall, *The Search for Human Chromosomes*,
DOI 10.1007/978-3-319-26336-6

Bacteriophage Virus which infects and replicates exclusively within bacteria, as an endoparasite. The T-phages have a genome which is double-stranded DNA. Upon introducing their DNA into the bacterial cell, they take over the internal chemistry to replicate millions of new phages. Eventually the bacterium bursts to release the newly formed phage particles.

Barr Body X chromosome that is condensed, heterochromatic and visible by staining the interphase nucleus. Present in females where there are two X chromosomes, but not in males with only one.

Camera Lucida A form of drawing aid which allowed the projected image from the microscope to be used as a guide while drawing the subject. The apparatus is adjusted so that the plane of the drawing paper and the microscope image can be observed simultaneously.

Centromere Primary constriction of a chromosome. The site of the kinetochore.

Chiasmatypie Term coined by F. A. Janssens for the phenomenon now called crossing over.

Chromatin Material which makes up the chromosome. It is composed of a DNA strand wrapped around histone proteins.

Colchicine A plant alkaloid generally found in species of autumn crocus of the genus *Colchicum*. Colchicine binds to tubulin, stopping production of microtubules and therefore the cell cycle at mitosis. By applying it to cell cultures and then washing it out, it is possible to create artificial polyploid cells.

Cotyledon Leaf, forming part of the embryo of seeds. Monocotyledons have a single one and dicotyledons have two. Gymnosperms have a variable number.

Cytoblastema Fluid described by Schwann from which he surmised the cell nucleus crystallised.

Dicentric Chromosome with two centromeres. There are no normal dicentric human chromosomes; they are all associated with chromosomal rearrangements.

Diploid Cell or organisms where each chromosome, with the exception of sex chromosomes, is present as pairs. This is normally described as $2n$.

Dominant Form of allele which is normally expressed in the heterozygous and homozygous condition.

Dosage Compensation A mechanism involved in balancing gene expression, primarily associated with the X chromosome, where overproduction in the homogametic sex (XX) compared with the heterogametic sex (XY) would cause disruption if it went unchecked. Dosage compensation of the X chromosome results in its condensation to form the Barr body.

Endoplasmic Reticulum Generally shortened to ER, the endoplasmic reticulum is a system of interconnected membranes within the cytoplasm of virtually all eukaryote cells. This seems to be a contiguous structure within the cell with various local features. It is important in maintaining the integrity of the cell and nucleus. Some areas have large numbers of ribosomes attached, this is rough ER, and the areas without ribosomes are smooth ER.

Endosperm Tissue surrounding the embryo in seeds. These are nutritive and frequently triploid.

Epigenetics The study of variation not directly attributable to genes and environmental influences affecting gene expression. It may or may not be inheritable, depending on the route the process takes. Cell differentiation into tissues and organs is an example of this.

Erythrocyte Red blood cell. These are cells with no nucleus in humans and other mammals.

Euchromatin DNA which makes up the genetically active component of the genome. It is not frequently used in its traditional sense of loosely compacted DNA as heterochromatic areas of the genome can be loosely compacted.

Eukaryote Organisms where cells contain a bounded nucleus containing the bulk of genetic material. Examples of eukaryotes include yeasts, algae, fungi and all higher organisms. The DNA is complexed with histones to make up chromatin.

Euploidy Polyploidy where the number of chromosomes is an exact multiple of the diploid set.

Exon Coding sequence in genes.

Fluorescent In Situ Hybridisation Usually referred to as FISH, this uses probes of complementary DNA to locate specific gene sequences which can be visualised by their fluorescence under ultraviolet light.

Fluorochrome A substance which gives fluorescence to a structure.

Fluorophore The chemical group which fluoresces in a fluorochrome.

Gene The unit of heredity. The term was introduced by a Danish geneticist, Wilhelm Johannsen, in 1909. This replaced the many other words which had been coined and discarded over the preceding years, such as pangens, gemmules, biophores and plasomes.

Genome Entire set of genes found in a cell. Functioning and nonfunctioning genes are included.

Genotype The genetic constitution of an organism. Only part of the genotype is expressed as the phenotype, and recessive genes in heterozygote form are not expressed.

Half-Life There are two meanings for this. Biologically it refers to the time for an organism to eliminate half of an administered dose of a drug. With reference to radioactive isotopes, this is the time needed to lose half of the original radioactivity due to decay into another non-radioactive isotope. With a half-life of, say, 10 days, this means every 10 days the radioactivity is halved, so after 20 days the radioactivity is 0.25 of the original amount when measurements started.

Haploid Cells or organisms where chromosomes are only present as single copies. This state is normally described as *n*.

Heterochromatin Condensed chromatin which is late replicating. Heterochromatin refers to the DNA and its functional attributes, rather than a description of its appearance. Consequently heterochromatin is frequently seen as heteropyknotic.

Heteropyknotic Condensed chromatin which stains darkly. Now referred to as positive heteropyknotic, it was originally heteropycnotic. The opposite is negatively heteropyknotic, originally isopycnotic, which gives a light staining chromosome. This term is most often used in reference to X chromosomes.

Heterozygote An organism that has inherited two forms of one gene. The individual will produce two types of allele, but may not be obviously heterozygous from the phenotype. Heterozygosity at one locus does not affect the possibility of homozygosity at another.

Hemizygous Having only one copy of a gene normally found as a pair. In mammals this is most often found in males in reference to genes on the X chromosome.

Histone Small proteins that bind to DNA. They contain large amounts of amino acids lysine and arginine. They form the protein core of DNA coiling which ultimately results in a visible chromosome. There are five main forms of histone, designated H1, H2A, H2B, H3 and H4. Their evolutionary conservation indicates their fundamental importance. Histone H4 is about 102 amino acids long, and between bovine and pea, there are only two differences.

Holometabolic Species where the adult form is arrived at by a process of metamorphosis, in insects from egg to several larval stages where growth takes place and then pupal transformation into the adult (imago).

Homologue One of a pair of homologous chromosomes, matching chromosomes found in diploid organisms.

Homozygous Two identical copies of a gene on different chromosomes. The individual can be heterozygous at other loci.

Homunculus A perfectly formed human encapsulated in a sperm head that grows when nourished in the womb. These were reportedly seen by some early microscopists.

Hydatidiform Mole A placental tumour which is normally androgenetic, that is, development from complete paternal set of chromosomes. The gynogenetic zygote forms an ovarian teratoma made up of poorly differentiated tissue types. These two states are important in observations of genetic imprinting.

Hyperdiploid A cell with more than the expected diploid number of chromosomes. This condition is normally found in neoplasias where marker chromosomes of unknown origin increase the total number of chromosomes.

Hypodiploid A cell having fewer than the expected diploid number of chromosomes. Note that this refers to the number. Fusions and rearrangements in neoplasias can result in a hypodiploid condition with the correct total amount of DNA.

Immunohistochemistry Process which uses molecule-specific antibodies tagged with an enzymatic system to locate an antigen which can then be visualised by a reaction involving the enzyme tag and deposition of coloured material at the site of interest.

Imprinting Epigenetic phenomena of gene expression and switching based on the parent of origin. The imprinted gene is the one which is silenced.

Intron Non-coding sequence found within a gene.

Karyokinesis A term used by Strasburger to describe what came to be known as mitosis, in plants.

Kinetochore Point of attachment of spindle fibres during cell division. The kinetochore is associated with the centromere.

Lectin Carbohydrate-binding protein which causes agglutination or precipitation of complex carbohydrates. These are naturally occurring compounds from a variety of sources, including bacteria, plants and fungi through invertebrate and up to mammalian cell membranes.

Lymphocyte Often referred to as white blood cells, lymphocytes are usually between 6 and 12 μ diameter. They are morphologically similar, but different types have different functions, so there are T-lymphocytes and B-lymphocytes. These cells are associated with immune function.

Lyon Hypothesis Suggestion by Mary Lyon that dosage compensation of the X chromosome can be maintained in cells by inactivation of all but one of the X chromosomes. In the normal condition of XX one would be inactivated, if there are three X chromosomes two would be inactivated, and so on.

Meiosis Cell division involved in the production of gametes. The chromosome number of a diploid cell (2N) is reduced by half to the haploid number N.

Microtubule The structural component of the cell nucleus which is responsible for movement of chromosomes. They are also found throughout the cell and are important in intracellular movement of molecules.

Mitogen Substance which stimulates mitosis in cells.

Mitosis Somatic cell division, distinct from division occurring in sex cells. Mitosis maintains the chromosome number in daughter cells and integrity of the chromosomes. It is arbitrarily divided into five broad descriptive stages: interphase, prophase, metaphase, anaphase and telophase.

Mutagen Chemical or physical agent, such as ionising radiation, which increases the mutation rate above the normally expected.

Morgan A measure of the relative distance between genes on the same chromosome. Crossover rates of 100 % equates to 1 Morgan. The more normal value is when the crossover rate is a few percent, and then it is expressed in centimorgans (cM).

Neoplasm A population of cells without normal control of cell division resulting in a tumour. These may be malignant or benign.

Nissl Granule Often called the Nissl body, these are ribonucleoprotein granules found in cell bodies and dendrites of neurones.

Nuclein The name given to the nucleic acid extract made by Miescher.

Nucleolus A specific part of the eukaryotic nucleus. It is visible in the interphase nucleus using a light or electron microscope, as a structural variation in the nucleus. It is the site of ribosome construction, and although a specific area, it does not seem specifically bound by a membrane.

Peptide A molecule made up of at least two amino acids joined by a peptide bond. Formation of a peptide bond requires joining two amino acids with the loss of a water molecule. Peptides come in various sizes so they can be dipeptides, tripeptides and, when large, polypeptides. Proteins are made up of one or more peptide chains.

Phenotype The outwards physical manifestation of an organism. This may be as obvious as hair colour or as obscured as blood groups. The phenotype is a result of the genotype and environmental influences.

Polyploid Individual having more than the diploid number of chromosome sets. There might be three complete sets in which case it would be described as triploid.

Polytene Chromosomes where there has been repeated rounds of DNA replication without separation of the resultant chromatids. Polytene chromosomes have been found widely in the animal kingdom, but are most commonly encountered in dipteran flies.

Proband The first affected family member through whom the family tree of inheritance is traced. This is the same as propositus.

Prokaryote Organisms lacking a nuclear membrane and containing a single strand of DNA which is not complexed with histone proteins. Bacteria are prokaryotes.

Propositus The first affected individual through whom a family tree is generated. This is the same as proband.

Purine An organic two ringed base. In nucleic acid these are adenine and guanine. The breakdown product of purines is uric acid.

Pyrimidine An organic single ringed base from nucleic acids. These are cytosine, thymine and uracil.

Recessive Characteristic where an allele is only expressed in the homozygous condition. The alternative homologue which is normally expressed is dominant.

Retinoblastoma A malignant childhood tumour of the retinal cells. It is autosomal dominant associated with chromosome 13 and is treatable with surgery.

Ribonucleoprotein A complex of RNA and protein, formed to protect RNA against premature degradation and help localise RNA within the cell.

Ribosome A small and structural particle which is the site of translation of messenger RNA into protein during protein synthesis.

Robertsonian Translocation Sometimes described as a whole-arm fusion. This is the end-to-end fusion of acrocentric chromosomes due to breakage and loss of the small arms.

Satellite In humans, part of a chromosome which is attached to the short arm of the chromosome by a stalk of variable length. The human satellite chromosomes are 13, 14, 15, 21 and 22. Not all satellites are visible in every cell, and satellite chromosomes can often be found associated together, apparently attached at the satellite end.

Spireme Term used by Flemming to describe the intertwined chromosome threads which were just visible and signified the start of cell division.

Statistical Errors Measurement errors which reduce precision by their random nature, but are for good reasons, assumed to sum to zero over repeated measurements. These errors can be accommodated by statistical techniques so as not to affect the overall accuracy.

Systematic Errors Errors which skew results in a particular direction resulting in lowered accuracy, but with no change in precision. Consistently over- or underestimating a result.

Telomere Terminal point of a chromosome. Associated with specific DNA repeat sequences.

Trait A particular phenotypic character of an individual. This may be obvious, such as hair colour, or hidden, such as a biochemical trait.

Transcription RNA synthesis from a DNA template. Catalysed by RNA polymerase, the polymerase recognises the promoter sequence and then starts assembling a chain of RNA with exact complementarity to the DNA sequence.

Translation Protein synthesis taking place in ribosomes using the transcribed RNA to produce a linear chain of amino acids. Post-translational changes to the chain activate the protein, and these can involve folding, cutting and splicing.

Transposon Sequence of DNA that usually consists coding and control regions which can be inserted or moved around within a genome.

Tubulin Generic term for a group of six different globular proteins. In genetics it generally refers to the two tubulin proteins which make up microtubules and have a molecular weight of about 50,000 Da.

Ukiyo-e 'Pictures of the floating world'. Prints and occasionally paintings made in Japan from the seventeenth to the nineteenth centuries. There were four highly skilled people involved in the production of a Ukiyo-e print. These were the designer who would be the named artist; the engraver who transfers the image to one or more wooden blocks, depending on the range of colours; the printer who applied the inks and paints to the blocks and transferred the image to handmade paper; and the publisher who coordinated the group and disseminated the finished work. During the nineteenth century, these products became immensely popular in Europe and influenced the impressionist painters.

X-ray Diffraction This is a method used to show the three-dimensional arrangement of atoms relative to each other in macromolecules such as DNA or proteins. Although it is possible to reconstruct the three-dimensional structure, the data comes in the form of a two-dimensional photographic image. It depends for its effectiveness on the wavelength of X-rays being about 0.154 nm, which is the same scale as interatomic distances.

Further Reading

Auerbach F (2013) The Zeiss works and the Carl-Zeiss Stiftung in Jena; their scientific, technical and sociological development and importance. HardPress Publishing. Reprint of 1904 original

Brenner S (2001) My life in science (Lives in science). Faculty of 1000 Ltd, London

Christie DA, Tansey EM (eds) (2006) Development of physics applied to medicine in the UK, 1945–90. Wellcome witnesses to twentieth century medicine, vol. 28. Wellcome Trust Centre for the History of Medicine at UCL, London. Freely available online at ww.ucl.ac.uk/histmed/publications/wellcome_witnesses_c20th_med

Crick F (1990) What mad pursuit: a personal view of scientific discovery. Basic Books, New York

Desmond A, Moore J (1991) Darwin. Michael Joseph, London

Edwards JH, Lyon MF, Southern EM (eds) The prevention and avoidance of genetic disease. Philos Trans R Soc Lond Ser B 319(1194):209–367

Fisher RA (1930) The genetical theory of natural selection. Clarendon, Oxford

Ford EHR (1973) Human chromosomes. Academic, London

Fraser Roberts JA (1940) An introduction to medical genetics. Oxford University Press, London

Gallagher, Dennis (eds) (2001) The human genome. Nature Palgrave, London

Garfield S (2013) Mauve: how one man invented a colour that changed the world. Faber and Faber Non Fiction, London

Garrod AE (1909) Inborn errors of metabolism. Henry Froude, London

Gribbin J (2002) Science: a history. Allen Lane/The Penguin Press, London

Griffiths P, Stotz K (2013) Genetics and philosophy: an introduction (Cambridge introductions to philosophy and biology). Cambridge University Press, Cambridge

Harper PS (2006) First years of human chromosomes. Scion, Oxford

Harper PS (2008) A short history of medical genetics. Oxford University Press, New York, NY

Kevles DJ (1995) In the name of eugenics: genetics and the use of human heredity. Harvard University Press, Cambridge, MA

McCarty M (1986) The transforming principle. W.W. Norton and Company, London

Perkin WH (2015) On the aniline or coal-tar colours. Andesite Press reprint of 1869 original, London

Peters JA (ed) (1959) Classic papers in genetics. Prentice-Hall, Englewood Cliffs, NJ

Seuanez HN (1969) The phylogeny of human chromosomes. Springer, Berlin

Sturtevant AH (1965) A history of genetics. Cold Spring Harbor Laboratory Press, Cold Spring Harbor, NY

Watson JD (1968) The double helix. Weidenfeld and Nicholson, London

© Springer International Publishing Switzerland 2016
W.J. Wall, *The Search for Human Chromosomes*,
DOI 10.1007/978-3-319-26336-6